Everyday Life-Environmentalism

This book provides one of the first systematic introductions to the Japanese concept of life-environmentalism, *Seikatsu Kankyō Shugi*. This concept emerged in the 1980s as a shared research framework among Japanese social scientists studying the adverse consequences of postwar industrialization on everyday life in communities.

Life-environmentalism offers a lens through which the agency of small communities in sustaining their everyday life and living environment can be understood. The book provides an overview of this approach, including intellectual backgrounds and foundational concepts, along with a variety of empirical case studies that examine environmental and sustainability issues in Japan and other parts of Asia. It also includes critical reflections on the approach in light of contemporary sustainability challenges. The empirical topics covered in the book include local community responses to development projects, resource governance, disaster response and recovery, and historical environmental preservation. The chapters are contributed by researchers working at the forefront of the field. It provides only a glimpse into the vast literature that awaits further exploration and engagement in the future.

The book is suitable for upper undergraduate students, graduate students, and researchers interested in environmental problems, sustainability and resilience, disaster mitigation and response, and regional development in Asian contexts, particularly Japan. It is well suited for courses in anthropology, geography, sociology, urban and regional planning, political science, Asian studies, and environmental studies.

Daisaku Yamamoto is Associate Professor of Geography and Asian Studies at Colgate University, USA. He holds a PhD in Geography from the University of Minnesota. His recent published works include: *Unravelling the Fukushima Disaster* (2016) and *Rebuilding Fukushima* (2017) (co-edited with Mitsuo Yamakawa); *Nuclear-to-Nature Land Conversion* (*Geographical Review*, 2020); *Cursed Forever? Exploring Socio-Economic Effects of Nuclear Power Plant Closures Across Nine Communities in the United States* (*Geoforum*, 2022).

Hiroyuki Torigoe is Professor at Otemae University, Japan, specialized in Environmental Sociology and Folk Cultural Studies. He holds a PhD from the Tokyo University of Education, and has taught at Kwansei Gakuin University, the University of Tsukuba, and Waseda University. He was the President of the Japan Sociology Society and the President of Otemae University. He is best known for his pioneering work on life-environmentalism based on his extensive fieldwork in Japan and overseas. He has published over 20 single-authored books, 17 co-authored or edited books, and numerous articles.

Everyday Life-Environmentalism
Community Sustainability and
Resilience in Asia

Edited by Daisaku Yamamoto
and Hiroyuki Torigoe

LONDON AND NEW YORK

First published 2024
by Routledge
4 Park Square, Milton Park, Abingdon, Oxon OX14 4RN

and by Routledge
605 Third Avenue, New York, NY 10158

Routledge is an imprint of the Taylor & Francis Group, an informa business

© 2024 selection and editorial matter, Daisaku Yamamoto and Hiroyuki Torigoe; individual chapters, the contributors

The right of Daisaku Yamamoto and Hiroyuki Torigoe to be identified as the authors of the editorial material, and of the authors for their individual chapters, has been asserted in accordance with sections 77 and 78 of the Copyright, Designs and Patents Act 1988.

All rights reserved. No part of this book may be reprinted or reproduced or utilised in any form or by any electronic, mechanical, or other means, now known or hereafter invented, including photocopying and recording, or in any information storage or retrieval system, without permission in writing from the publishers.

Trademark notice: Product or corporate names may be trademarks or registered trademarks, and are used only for identification and explanation without intent to infringe.

British Library Cataloguing-in-Publication Data
A catalogue record for this book is available from the British Library

ISBN: 978-1-032-02751-7 (hbk)
ISBN: 978-1-032-02752-4 (pbk)
ISBN: 978-1-003-18503-1 (ebk)

DOI: 10.4324/9781003185031

Typeset in Times New Roman
by SPi Technologies India Pvt Ltd (Straive)

Contents

List of Figures viii
List of Table x
List of Contributors xi
Preface and Acknowledgments xv

1 Introduction 1
  DAISAKU YAMAMOTO

2 Theorizing Everyday Life: The Life-Environmentalist Way 25
  DAISAKU YAMAMOTO

PART I
Developmental Impulse and Everyday-Life Organizations 43

3 Local Rules: Sustaining Local Everyday Life with
  Aqua-Tourism 45
  TAKEHITO NODA

4 Coexistence Without Consensus: The Role of a Life
  Organization in Mediating between Fishermen and
  Surfers in a Coastal Community 58
  SHUSUKE MURATA

5 When Civil Society Falls Short: Rural Community
  Response to a Resort Development Project 74
  DAISAKU YAMAMOTO AND YUMIKO YAMAMOTO

PART II
Governing Everyday-Life Spaces   **89**

6   From Dichotomous Interpretations to Spectrum Thinking: Formation of a Community Organization in a Nuclear Host Locality   91
ATSUSHI YAMAMURO

7   "Public" (*gong*) as Village Norm: Urbanization and Community Response in China   105
MEIFANG YAN

8   Multilayered Commons Space: Dry Riverbed Use in a Local Community in Ibaraki, Japan   118
TAKAAKI ISOGAWA

PART III
Living with Disasters   **129**

9   Why Do Victims of the Tsunami Return to the Coast?   131
KYOKO UEDA AND HIROYUKI TORIGOE

10   The Roots of Resilience: Forest Commons and the Cultivation and Disappearance of Livelihood Security in a Nuclear Disaster–Afflicted Community   142
HIROYUKI KANEKO

11   Apparitions and the Recovery of Livelihoods after the 2011 Tōhoku Earthquake and Tsunami Disaster   156
KIYOSHI KANEBISHI

PART IV
Historic Environment and Urban Communities   **167**

12   Living Traditional Culture: Gujo Dance in Hachiman Town, Gujo City, Gifu Prefecture, Japan   169
SHIGEKAZU ADACHI

13 Embracing the Enemy's Legacy: Historical Environmental
    Preservation in Daegu, South Korea   181
    RIE MATSUI

14 Boxing Camp as a Community School: Local Boxers
    in Metro Manila, Philippines   191
    TOMONORI ISHIOKA

PART V
Critical Reflections and Prospects   **205**

15 Empirically Speaking: Life-Environmentalism,
    Environmental Justice, and Feminist Political Ecology   207
    DAISAKU YAMAMOTO, SOPHIA FERRERO AND KEEGAN KESSLER

16 Life-Environmentalism, Critiques, and Prospects:
    Focusing on the Experientialist Approach   224
    YASUSHI ARAKAWA

17 The Future of Life-Environmentalism: A Sympathetic
    Critique   235
    MASAHARU MATSUMURA

PART VI
**Translated Excerpts from the *Sociological Theory of Environmental
Problems* (1989)**   **253**

18 Original Introduction of Life-Environmentalism (1989)   255
    HIROYUKI TORIGOE

    Index   284

Figures

1.1	Estimated global population distribution as of 1500	5
1.2	Countries that have been under European control	5
1.3	Contemporary political regimes of the world (2019)	6
1.4	Per capita GDP growth since take-off	7
1.5	Number of articles with the term appearing in the abstract of peer-reviewed articles in English-language scholarly journals	13
1.6	Number of articles with the term appearing in the title of Japanese-language academic publications (including books and journals)	14
2.1	Positioning of life-environmentalism. Axes imply desirable nature–human relations	26
2.2	Conceptual flow leading to changes in living environment	38
3.1	Water area where residents and tourists co-use	46
3.2	Signboard for tourists in the Gion District of Kyoto	47
3.3	Water area for tourists in front of the JR Matsumoto station	48
3.4	*Kabata* area in the Harie District	52
4.1	Location of Kamogawa	61
4.2	Kamogawa fishing port and Ōura district on the higher ground	62
4.3	Set net fishing	63
4.4	Locations of Ichinori and Akatei in 1977	64
4.5	Kamogawa Fishing Port after the development of the Kamogawa Marina (January 2003)	65
4.6	Age compositions of *Teichi* members	67
4.7	Migrant surfers, also members of *Teichi*, carrying the local deity of Ōura	71
5.1	Matsukawa Village, Nagano	78
5.2	Gōdohara alluvial fan area on the outskirts of the Japan Alps (Hida Mountains)	78
6.1	Location of Tōkai Village and JCO	95
6.2	Four neighborhood associations of the areas adjacent to JCO	100
7.1	Schematic map of Village X and a street in the village	109
7.2	Rebuilt *Guandimiao* temple, (left), entrance of the temple (middle), and inside of the temple (right)	110

8.1	Community X and its land use	121
8.2	Vegetable gardens in *kawabata*	122
8.3	Drying clothes in *kawabata*	125
9.1	Moune community in Kesennuma City, Miyagi Prefecture	135
10.1	Evacuation designations in the area around Fukushima Daiichi as of September 2014	143
10.2	Locations of chestnut forests	148
10.3	Changes in forest ownerships	151
11.1	Notes written on a desk found in a tsunami-hit area	157
11.2	Schematic diagram of the positioning of the "living deceased"	162
12.1	A *yakata* float at the Gujo Odori festival	171
12.2	Two concepts of "past"	177
12.3	*Kiriko* lantern	178
13.1	Bukseongro neighborhood in Daegu	183
13.2	Bukseongro tool museum	188
14.1	Squatter area near the boxing gym	193
14.2	Dining room in the gym	198
18.1	Process of forming a specific action based on experience	272
18.2	Process leading to changes in living environment	276

Table

18.1   Patterns of Decision-Making in Local Communities   273

Contributors

Shigekazu Adachi is Professor in the Department of Sociology, Otemon Gakuin University, Japan. He holds a PhD in Sociology from Kwansei Gakuin University. His research focuses on folk culture, cultural tourism, and regional development. His published work includes: "The Interaction Between Humans and Nature: An Investigation into Coexistence" (*Journal of Environmental Sociology*, 2017: In Japanese) and "Toward fieldwork for a community's sense of everyday life: A case study of Hachiman Town, Gifu Prefecture" (*Advances in Social Research*, 2008: In Japanese).

Yasushi Arakawa is Professor in the Department of Human Science, Taisho University, Japan. He holds a PhD in Sociology from the University of Tsukuba. His research interests include the use and meaning of public space in Japan and China, and forest commons and conservation of resident livelihoods. His published works include "Today's implications of life-environmentalism: Focusing on sustainable society" (*The Journal of Studies in Contemporary Sociological Theory*, 2009: In Japanese) and "Have Parks Enriched Urban Environments?" (*Environmental Sociology: Twelve Perspectives to Analyze Living*, 2019: In Japanese).

Sophia Ferrero is a recent graduate from Colgate University, USA, with a bachelor's degree in Environmental Geography and Asian Studies. Her research interests lie in using a localized approach to understand how and why communities conceptualize and respond to environmental issues and projects that utilize community-based GIS in environmental problem-solving.

Tomonori Ishioka is Professor in the Department of Sociology, Nihon University, Japan. He holds a PhD from the University of Tsukuba. His works include: *The Bottom Worker in East Asia: Composition and Transformation under Neoliberal Globalization* (co-edited with Hideo Aoki, 2023), "The Habitus Without Habitat: The Disconnect Caused by Uprooting During Gentrification in Metro Manila" (*Social Theory and Dynamics*, 2016), and "Boxing, Poverty, Foreseeability: An Ethnographic Account of Local Boxers in Metro Manila, Philippines" (*Asia Pacific Journal of Sport and Social Science*, 2013).

Takaaki Isogawa is Instructor in the Department of Sociology, Shitennoji University, Japan. He holds a PhD in Sociology from the University of Tsukuba. His research interests include the relationship between waterways and local livelihoods. His publications include: "A Review of Environmental Sociology in Japan from the Perspective of Life-environmentalism" (*Research reports, University of Hyogo*, 2008: In Japanese) and "Devastation of Rural Community Space and Rural Studies: How to Approach Neglected Graves, Vacant Houses and Abandonment of Farming" (*Journal of Rural Studies*, 2016: In Japanese).

Kiyoshi Kanebishi is Professor in the School of Sociology, Kwansei Gakuin University, Hyogo, Japan. He holds a PhD from Kwansei Gakuin University. His published works include: *Yobisamaseru Reisei-no Shinsaigaku* [Studying the Awakened Spirituality of the 2011 Tōhoku Disaster] (2016, In Japanese) and "The Inner Shock Doctrine: Life Strategies for Resisting the Second Tsunami" (*Social Theory and Dynamics*, 2016), *Life-Environmentalism and Community Research* (2018: In Japanese).

Hiroyuki Kaneko is Associate Professor in the Department of History, Tohoku Gakuin University, Japan. He holds a PhD in Human Sciences from Waseda University. His publications include: "Radioactive Contamination of Forests and RuralCommunities: History of Relations between Upland Forested Areas and the Village in Kawauchi Village, Fukushima Prefecture"(*Journal of Forest Economics*, 2022: In Japanese) and "Radioactive Contamination of Forest Commons: Impairment of Minor Subsistence Practices as an Overlooked Obstacle to Recovery in the Evacuated Areas" (*Unravelling the Fukushima Disaster*, 2016).

Keegan Kessler is a recent graduate from Colgate University, USA, with a bachelor's degree in Geography and Japanese. She is from the island of Kaua'i in Hawai'i, and as a Native Hawaiian and Asian scholar, she is interested in critical human-environment research with particular focus on Indigenous and AAPI (Asian American and Pacific Islander) experiences.

Masaharu Matsumura is the director of the nonprofit research organization NORA (https://nora-yokohama.org), Japan. He was Associate Professor in the Faculty of Human and Social Studies in Keisen University until 2020. He holds a MA from the Tokyo Institute of Technology. His works include: "The Satoyama Movement and Its Adaptability: Beyond Ideology and Institutionalization" (*Adaptive Participatory Environmental Governance in Japan: Local Experiences, Global Lessons*, 2022) and "Creating a Sustainable Community Based on Nearby Nature: Suburban Satoyama Governance History in the Heisei Era" (*Journal of Environmental Sociology*, 2018: In Japanese).

Rie Matsui is Associate Professor in the Department of Community Design, Atomi University, Tokyo, Japan. She has a PhD from the University of Tsukuba. Her interests include Postcolonial Studies in Architecture and

Ethnographic/Archival Research on the contemporary use of colonial Japanese-style houses in Korea. Her publications include: "Urban Regeneration Through Landscape Preservation Movement: From a Case Study of Renovation Planning of Historical Architectures in Daegu, South Korea" (*Contemporary Sociological Studies*, 2017: In Japanese).

Shusuke Murata is Professor in the Faculty of Regional Sciences, Tottori University, Japan. He received his PhD from the University of Tsukuba. His research interests include Sport Tourism and Community Development, with particular interest in fishing communities. He is the author of *Sustainable Sport Tourism and Spatial Conflicts* (2017: In Japanese) and "New Actors in Rural Communities" (book chapter in *Introduction to Regional Policy Studies*, 2019: In Japanese).

Takehito Noda is Associate Professor in the Faculty of Social Policy and Administration, Hosei University, Japan. He holds a PhD from Waseda University. Trained as an environmental sociologist, his current research projects include community-centered aqua-tourism and post-disaster local development. His publications include: *Local Revitalization Starting by Wells: Disaster Prevention and Tourism From the Perspectives of Everyday Life* (2023, In Japanese) and "Why Do Local Residents Continue to Use Potentially Contaminated Stream Water After the Nuclear Accident?" (*Rebuilding Fukushima*, 2017).

Hiroyuki Torigoe is Professor at Otemae University, Japan, specialized in environmental sociology and folk cultural studies. He holds a PhD from the Tokyo University of Education, and has taught at Kwansei Gakuin University, the University of Tsukuba, and Waseda University. He was the President of the Japan Sociology Society and the President of Otemae University. He is best known for his pioneering work on life-environmentalism based on his extensive fieldwork in Japan and overseas. He has published over 20 single-authored books, 17 co-authored or edited books, and numerous articles.

Kyoko Ueda was Professor in the Department of Sociology, Sophia University, Japan. She had a PhD, University of Tsukuba, and MA in Comparative Education from the University of London. Her published work includes *Villages at the Crossroad of Existence* (2016: In Japanese), "How Does Evacuation Cause the End of Livestock Farming?: The Fukushima Dai-ichi Nuclear Power Plant Accident and the Evacuation of Livestock Farmers" (*Journal of Environmental Sociology*, 2019), and "Why Do Victims of the Tsunami Return to the Coast?" (*International Journal of Japanese Sociology*, 2012).

Daisaku Yamamoto is Associate Professor of Geography and Asian Studies at Colgate University, USA. He holds a PhD in Geography from the University of Minnesota. His recent published works include: *Unravelling the Fukushima Disaster* (2016) and *Rebuilding Fukushima* (2017) (co-edited with Mitsuo Yamakawa); "Nuclear-to-Nature Land Conversion" (*Geographical Review*, 2020); and "Cursed Forever? Exploring Socio-Economic Effects of Nuclear Power Plant Closures Across Nine Communities in the United States" (*Geoforum*, 2022).

Yumiko Yamamoto is Lecturer in the University Studies and a research affiliate in the Department of Geography, Colgate University, USA. She holds a doctoral degree in civil engineering from the University of Tokyo. Her interests include alternative economies and community development, resident movements and recreational development, and remote sensing in rural land-use research. Her works include "Community Resilience to a Developmental Shock: A Case Study of a Rural Village in Nagano, Japan" (*Resilience: International Policies, Practices and Discourses*, 2013).

Atsushi Yamamuro is Professor in the Graduate School of Regional Resource Management, Hyogo Prefectural University, Japan. He has a PhD in Sociology from Kwansei Gakuin University. His research interests include nuclear energy facilities and local host communities. His published works include: "Hesitation Among Citizens of Nuclear Host Communities: A Provisional Approach to Understanding Their Continuous Efforts to Maintain Security of Life" (*Journal of Environmental Sociology*, 2012: In Japanese).

Meifang Yan is Lecturer at the Faculty of Sociology, Ryukoku University, Japan. She holds a PhD from Waseda University. Her work includes: *Studies of Village Order in Japan and China: Public Village Matters form a Daily Life Theory Perspective* (2021: In Japanese) and "Cooperativity and Public Village Matters in the Villages of China: From a Case Study of X Village Under Village-urbanization" (*Japanese Sociological Review*, 2017: In Japanese).

Preface and Acknowledgments

When one of the editors (Yamamoto) first offered *Sustainable Livelihoods in Asia*, an undergraduate course in Geography and Asian Studies at Colgate University in 2014, the intention was to explore various socio-environmental issues in Asia, with a particular focus on Japan. Drawing primarily from the English-language geography and related literature, the course began by discussing basic geographic concepts such as space and place, referencing renowned Anglophone geographers like Doreen Massey, David Harvey, and Edward Soja. Key theories and ideas on human–environment relationships were then introduced, encompassing environmental justice, sustainable livelihoods, common pool resources, and resilience. These were followed by discussions of local case studies that examine regions and communities facing disasters, environmental degradation, or external development pressures.

However, the realization that "Asia needs to be taken more seriously" became clear after the initial offering of the course. Most of the theories and concepts on human–environment relationships addressed in the course were developed within the historical and geographical contexts specific to Euro-American societies, without critical reflection on the potential consequences of applying them to Asia. This led to the emerging question: What do theories grounded in the concrete realities of Asian places look like?

Prior to teaching the course, Yamamoto was already familiar with *seikatsu Kankyō shugi*, or life-environmentalism, which serves as the central focus of this book. However, there were limited English-language materials suitable for systematic introduction in college-level courses. Accordingly, students were occasionally presented with brief summaries and takeaways of life-environmentalism studies that were written in Japanese. Nonetheless, it would have been more beneficial for them to read and critically evaluate the materials on their own, rather than relying on the instructor's interpretations.

When Yamamoto shared his frustration with his lifelong mentor, Kazunori Matsumura, the idea of producing a book on life-environmentalism, or the broader *seikatsu-ron* (生活論: literally, "life-approach"), was suggested. Furthermore, he encouraged me to collaborate with Hiroyuki Torigoe, co-editor of this book. While Yamamoto was already acquainted with Torigoe, he had not though about working with one of the pioneers in the field. Therefore, it was this encouragement that ultimately made the realization of this book project possible.

In planning for the book, we decided that the main focus should be on empirical case studies of life-environmentalism, rather than theoretical expositions and discussions. This decision aligns with the original motivation of the book, which is to make concrete case studies available to an English-language audience, particularly undergraduate and early graduate students. This orientation also reflects the spirit of life-environmentalism and *seikatsu-ron*, which emphasize the importance of empirical studies based on intensive fieldwork and see theory as valuable to the extent that it can inform field-based research.

For readers familiar with environmental sociology and its related fields in Japan, the content of this book should be hardly novel. The concept of life-environmentalism and its intellectual foundation, *seikatsu-ron*, are well-known terms, and some of the empirical case studies in this book have been already published in Japanese. We are also aware that there are various criticisms of life-environmentalism. However, considering the limited number of English-language publications on life-environmentalism, we believe it is valuable to first focus on introducing this approach and its empirical studies to broader international audiences before engaging in deep self-reflection and critical debates.

That being said, the book does include introductory chapters that contextualize life-environmentalism research within broader contexts and explain key concepts of life-environmentalism. It also features final chapters that reflect on contributions of life-environmentalism and debates surrounding it, even if still preliminarily. In addition, the appendix includes one of Torigoe's earliest texts on life-environmentalism, originally published in 1989, which laid out many of its foundational ideas. To preserve the nuances of the original Japanese texts, the translation maintains some ambiguous expressions, inviting readers to contemplate and interrogate them further.

Most of the chapters in this book are written by what can be considered the second generation of life-environmentalism or *seikatsu-ron* scholars, some of whom were graduate students of Hiroyuki Torigoe. While many of the authors have a background in sociology, their scholarship, rooted in intensive fieldwork in local communities, shares common ground with field-oriented researchers in other disciplines such as anthropology, geography, and environmental studies. Therefore, we hope that this book will be read widely across disciplinary and geographical boundaries.

At the same time, we wish to emphasize the importance of engaging with Japanese-language texts to fully grasp the intellectual roots, scope, and depth of life-environmentalism and *seikatsu-ron* research. Much of the extensive and diverse literature on *seikatsu-ron*, which forms the foundation of life-environmentalism, has yet to be translated into English or other languages. This book provides only a glimpse into the vast literature that awaits further exploration and engagement in the future.

We recognize that the book's title, "Everyday Life-Environmentalism: Community Sustainability and Resilience *in Asia*," may appear overly ambitious, given that the majority of case studies are centered on Japan, with only three chapters exploring non-Japanese regional contexts. We want to make it clear that we do not intend to assert universal applicability of the life-environmentalism approach across all of Asia. Conversely, we also see no compelling reason to believe that the fundamental tenets of life-environmentalism are exclusively valid in Japan. Indeed, our hope for this book is to invite critical discussions and prompt alternative theorization and empirical insights rooted in diverse sociocultural contexts.

This book would not have come to fruition without the fortuitous encounters and the support of numerous individuals. As mentioned earlier, Yamamoto is particularly indebted to Kazunori Matsumura, a rural sociologist who has consistently emphasized the significance of scholarship that remains closely connected to people's everyday lives and warned against excessive abstraction. Matsumura has also been instrumental in Yamamoto's continued engagement with Japanese environmental sociology, despite his institutional training in the field of human geography in North America.

Colgate University, a small liberal arts college in Upstate New York where Yamamoto is currently jointly appointed in the Geography Department and Asian Studies Program, has provided the freedom and encouragement to undertake a project that transcends traditional disciplinary boundaries. It has served as a space where several crucial encounters have taken place. First and foremost, it was the students who served as the true inspiration for this book project. We received invaluable comments, questions, and critiques from them as readers of some of the earlier manuscripts. In addition, Clare Edminster and Jordan Shapiro provided critical assistance during the translation, editing, and proofreading process. Sophia Ferrero and Keegan Kessler, co-authors of a chapter in this book, also offered perspectives as students with academic training rooted in the U.S. academic institution.

We are also immensely grateful to Alex Sklyar for his invaluable assistance with the translation and revision of several chapter manuscripts. As a cultural anthropologist who has conducted extensive ethnographic work in Japan, his insights have undoubtedly enriched the quality of the text. We would also like to express our appreciation to Bill Meyer who generously dedicated his time to read portions of the book manuscript and provided us with tremendously useful comments and suggestions, Paul Plummer for his encouraging comments and critical feedback, and Jay Bolthouse and Glen McCabe who translated some of the manuscripts that were originally written in Japanese. Additionally, we would like to acknowledge the Asian Studies Program, along with other divisions of Colgate University, for providing us with numerous opportunities to engage with scholars of Asia, including Prasenjit Duara and David Haberman. These encounters have instilled in us the confidence to voice more thorough and critical perspectives on environmental issues and to assert knowledge grounded in Asia.

As the completion of the book manuscript was approaching, Yamamoto visited Minneapolis to participate in a two-day panel session celebrating the retirement of Professors Eric Sheppard and Helga Leitner, from whom he had studied as a graduate student many years ago. While the content of this book may not directly align with his research during that time, the approach adopted in producing this book—learning together with students and humbly from those who possess deep knowledge—clearly reflects their teaching and research style. For this, a deep appreciation goes to them.

Finally, this book is dedicated to Kyoko Ueda, a promising scholar and a wonderful person, whose contribution of a newly written chapter to the book was anticipated with great enthusiasm. Her untimely passing is a profound loss to the field of life-environmentalism and beyond that will be deeply felt.

1 Introduction

Daisaku Yamamoto

What comes to your mind when you think of "studying environmental or sustainability issues in Asia?" Some may think of climate change and its impact on different parts of Asia, while others may think of the images of serious air pollution in China, deforestation in Indonesian rainforests, or nuclear fallout in Japan. Some others may think of the prospects of clean energy and environmental technologies in Asia. These images and knowledge of challenges to sustainability in Asia are generated and solidified through mass media and social media, and through primary, secondary, and even postsecondary education. At the core of this common-sense knowledge are underlying assumptions such as:

- Climate change is the most urgent problem of the day in Asia, just like everywhere else.
- Scientific analysis offers objective knowledge necessary for environmental sustainability.
- The nation state is the most effective scale of sustainability policies and institutions, promoting, for example, renewable energy technologies, recycling programs, or sustainable city designs.
- Civil society and social movements are the key to promote environmental justice.
- We must protect nature from harmful human interventions.
- We must educate people who lack advanced knowledge, technologies, and capacities to achieve sustainable development.

This book may disappoint those who firmly believe that these statements are indisputable facts, as I consider them untested assumptions that could hinder our understanding and recommendations for various sustainability and socio-environmental challenges in Asia. In fact, there is an increasing awareness that theories and discourses of sustainability popular in industrialized, Euro-American countries may not always accurately capture the nature of sustainability challenges and provide useful solutions in other geographical contexts (Duara 2014; Muldavin 2008).

For example, Duara (2014) argues that while environmentalism in heavily industrialized regions often focuses on wilderness protection and certain lifestyle changes, in other parts of the world, "conserving the environment means conserving the sources of their livelihood" which are threatened by industrialization and urbanization (Duara 2014, 40). Moreover, concepts such as democracy, civil society, and justice may not hold the same significance, public acceptance, and functionality in non-Western societies, where "the values to which struggles for environmental justice must appeal are still strongly shaped by the (upper) middle classes" (Williams and Mawdsley 2006, 667). Various efforts have been made to critically adapt these theories and discourses to the realities of sustainability problems in non-Euro-American contexts (e.g., Yeh, 2009; Williams and Mawdsley, 2006). The growing interest in indigenous studies is also indicative of this broader trend (Coombes et al. 2012, 2013, 2014).

However, approaches primarily developed in Western, particularly English-speaking scholarship, still dominate the field of environmental studies and related disciplines. While these contributions are undoubtedly significant and provide powerful analytical perspectives in diverse situations, there is much to be learned from bodies of literature that have emerged more organically within non-English-speaking scholarship.

Life-environmentalism: A Brief Background

In this book, we draw on one specific body of research primarily developed in Japan known as *Seikatsu Kankyō Shugi*, which we refer to as *life-environmentalism*.[1] The origins of life-environmentalism can be traced to a research project conducted in the areas surrounding Lake Biwa in the late 1970s to the 1980s. During the post–World War II period, the Pacific Manufacturing Belt underwent intensive industrialization and urbanization, leading to resource exploitation, the establishment of energy infrastructure, and the migration of labor from peripheral areas. Numerous regions across the country were experiencing rapid transformations, and Lake Biwa, situated in the upstream section of the Yodo River System that runs through Kyoto and Osaka, was one of them.

Traditionally, the lake has served as a natural reservoir for flood control and a water source for downstream regions (Kawanabe, Nishino, and Maehata 2012; Kada 2006). In 1972, the Lake Biwa Comprehensive Development Plan, similar to many state-driven projects elsewhere in the country, was initiated to enhance and secure water access for cities like Osaka and Kobe. It involved transforming Lake Biwa through the construction of a multipurpose dam with walls and a floodgate, and extensive embankment of the shores and streams of the lake region (Kada 2006). Combined with the industrialization of agriculture and the modernization of lifestyles, it began to significantly impact the livelihoods of residents in the vicinity of the lake. Effects included a decline in the freshwater fishery, water pollution from agricultural runoff and household wastewater, and the abandonment of the use of lake water for daily purposes.

A group of researchers in a variety of fields, including sociology, anthropology, geography, and folklore studies (*minzokugaku*), with Hiroyuki Torigoe as the project leader, conducted a seven-year-long research project in communities surrounding Lake Biwa. Their work culminated in two edited volumes: *Mizu to Hito no Kankyōshi—Biwako Hōkokusho* (Environmental History of Water and People: Lake Biwa Report) (Torigoe and Kada 1984) and *Kankyō Mondai no Shakai Riron—Seikatsu Kankyō Shugi no Tachiba Kara* (Social Theory of Environmental Problems: From the Standpoint of Life-environmentalism) (Torigoe 1989). Since then, Torigoe, his colleagues, and their students have produced a variety of empirical studies on a wide range of socio-environmental issues, establishing what may be considered a school of thought within the field of environmental sociology and its related disciplines in Japan.

In essence, the perspectives of life-environmentalism, in contrast to the oft-held views on contemporary sustainability issues as described above, prompt us to focus on and highlight the importance of:

- Securing and improving people's lives and livelihoods.
- "Non-scientific" knowledge as a practical means to enhance sustainability.
- Locally specific, community-level knowledge, institutions, and technologies.
- The roles of folk culture, traditions, and "pre-modern" social organizations in securing people's well-being.
- The use of "nature" as a potentially viable means of conservation and sustainability.
- Active agency of rural communities and marginalized populations.

These points of arguments are most definitely not unique to life-environmentalism; many of these arguments are made in a growing literature pertaining to sustainability in the Euro-American scholarship. Nevertheless, it is worthwhile to pay attention to life-environmentalism because of its unique ways of theorizing based on a distinct epistemology and ontology that are culturally grounded in Japan (Chapter 2), and because of its concrete empirical insights that appear to resonate well with the minds of those whom it examines. Empirical studies informed by and theoretical perspectives embodied in life-environmentalism have a potential to enrich scholarly engagements on such issues as sustainable development, community resilience, and environmental justice.

Proponents of life-environmentalism have emphasized from the start a *field-workers'* theory, rather than being a theoretical framework generated from an extensive literature survey or philosophical reflection. Life-environmentalism did not emerge primarily as a reaction to the intellectual developments in Euro-American scholarship at the time, nor did it claim to have broad applicability far beyond its original empirical case study sites. However, there is now a considerable body of research that has been influenced, at least in part, by life-environmentalism, extending beyond the initial scope of the Lake Biwa research. This presents an opportune moment to situate life-environmentalism within broader historical, geographical, and theoretical contexts. Therefore, the purpose of this introductory chapter is to retrospectively frame life-environmentalism and articulate its approach for the English-language audience.

First, I provide a brief contextual background on "Asia," the setting in which life-environmentalism was developed and where most of its empirical studies have been conducted to date. I then discuss the intellectual context of life-environmentalism, comparing and contrasting it with several socio-environmental theories that are relatively well known in the English-language literature and that originate from the industrialized West. I highlight potential challenges in applying these theories to study various socio-environmental issues in non-Euro-American contexts and suggest the potential usefulness of life-environmentalism as an alternative approach. Finally, I outline the structure of the book.

Historical and Geographical Contexts

Torigoe (2014) states that life-environmentalism finds its greatest applicability in densely populated areas like Japan and China, where "nature mingles with everyday life and streams and hills are part of human habitation, not a wilderness" (22), and especially where "local communities are strong" (22). Additionally, Torigoe hints at the potential applicability of life-environmentalism in Southeast Asia and South America, although he does not discuss its geographical limitations at length. In this chapter, I aim to explore the historical and geographic contexts in which life-environmentalism emerged, in order to provide a better situational understanding of this approach.

Cursory Overview of Asia

Asia is home to approximately 60% of the world's population (7.7 billion), with China and India alone accounting for 2.8 billion people. What is more significant is that sizable portions of Asia have been *densely populated for a long time*, pre-dating European colonization and modernization (Figure 1.1). This suggests that people had relatively successfully developed ways of sustaining a relatively large population through practices of "living off the land." Various factors, such as favorable environmental conditions, production techniques, social organizations, or political institutions, may have contributed to this success. These factors also relate to one of the characteristics, both advantageous and potentially limiting, of life-environmentalism. That is, life-environmentalist perspectives may not provide a particularly useful framework in a new settler society, where a group of people (possibly heterogeneous in their origins) with few ties to the land live their lives, employing modern technologies and institutions form the start.

Asia is not only densely populated but also remarkably diverse. A testament to this cultural diversity is the existence of over 2,300 languages in Asia, whereas Europe has fewer than 300 (Ethnologue 2023). Additionally, while Africa, the Pacific Islands, and Latin America also boast a significant number of living languages (2,158, 1,324, and 1,064, respectively), it is important to

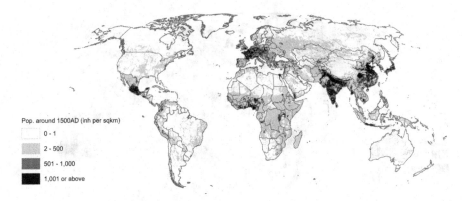

Figure 1.1 Estimated global population distribution as of 1500. Created by the author based on Goldewijk (2022).

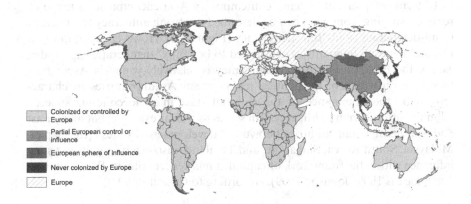

Figure 1.2 Countries that have been under European control. Created by the author based on Fisher (2015) and other sources.

note that European languages are not currently the official or dominant languages in many Asian countries. This factor may hinder the smooth flow of ideas, as well as concurrent social actions and movements, within the region compared to, for example, Latin America. The linguistic diversity in Asia reflects, to some extent, the varied colonial experiences that the region has undergone (Figure 1.2). While certain parts of Asia have been colonized by European powers, other areas have only experienced partial control (e.g., China), or have never been formally colonized by European powers (e.g., Thailand and Korea). Additionally, Japan stands as a unique case as both a former colonizer and a country within Asia. This cultural and historical diversity has the potential to complicate postcolonial discourses in Asia. It is questionable, at the very least, that one can characterize Asia solely as a "colonized" region.

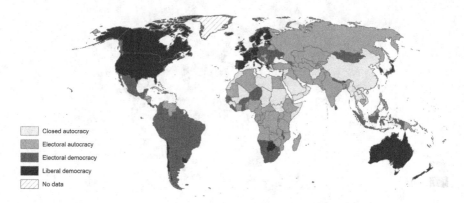

Figure 1.3 Contemporary political regimes of the world (2019). Created by the author based on Our World in Data (2023).

In terms of political systems, contemporary Asia encompasses a range of regimes, varying from what can be described as strong autocracy to anocracy (semi-democracy) and "full" democracy (Figure 1.3). However, it is important to note that even in countries considered to be "fully" democratic (e.g., India, Japan, the Philippines), formal democracy is relatively young in Asia.[2] Furthermore, regardless of the political regime, many Asian countries are characterized by the dominance of the national state in socioeconomic spheres. China's authoritarian political system serves as a clear example, but many East and Southeast Asian countries known as "developmental states" (e.g., Japan, Malaysia, South Korea, Singapore, and Thailand) also exhibit significant state influence within the framework of capitalist market economies (Woo-Cumings 1999; Deans 1999; Johnson 1999). According to Castells (1992),

> a state is developmental when it establishes as its principle of legitimacy its ability to promote and sustain development, understanding development as the combination of steady high rates of economic growth and structural change in the productive system, both domestically and in its relationship with the international economy.
>
> (56–57)

However, this does not imply that Asian states (or any states, for that matter) are always successful in achieving their goals. For example, Rudolph and Rudolph (1987) describe India as a "strong-weak state," characterized by a powerful and ambitious, yet not always effective, state apparatus. The crucial point here is that, regardless of the styles or "degrees" of democracy, an authoritarian state is a defining feature of political systems in most Asian countries. For our purposes, this signifies that local communities in Asia consistently confront an ambitious (albeit not always capable) state that tries to impose its will, often through various developmentalist or modernizing projects, frequently in the name of the greater national good.

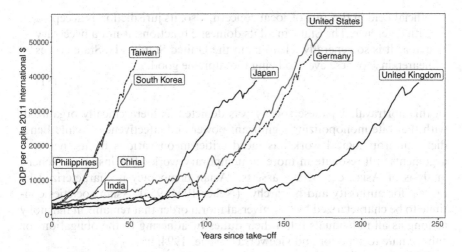

Figure 1.4 Per capita GDP growth since takeoff.
Sources: Bolt and van Zanden (2020); Pakrashi and Frijters (2017).

Compared to major industrialized countries in Europe and America, Asian countries have undergone a much faster sequence of economic growth, sectoral transformation, and even deindustrialization (Whittaker et al. 2020) (Figure 1.4). Additionally, demographic transitions have accelerated, leading to "premature aging," particularly in East Asia (Whittaker et al. 2020). One important implication of this "compressed" modernization is that "pre-modern" aspects of society may coexist and be visible alongside modern institutions and technologies.

In summary, a long-standing dense population, historical-cultural diversity, authoritarian regimes, and compressed modernization are the fundamental characteristics of many parts of Asia. These factors provide the broader context in which life-environmentalism developed.

Modernizing State and "Pre-modern" Realms of Everyday Life

The above characterization, though overly simplified, serves as a crucial baseline assumption for life-environmentalism. For instance, the coexistence of "pre-modern" customs and practices of local communities with the realm of the modern state and its institutions has been often disregarded or misunderstood by observers. In her seminal anthropological work on Japan, Ruth Benedict (1946) made the following observation about modernizing Japan since the Meiji Restoration:

> The true difference between the Japanese form of government and such cases in Western Europe lies not in form but in functioning.... At the topmost level of policy "popular opinion" is out of place. The government asks only "popular support." When the State stakes out its own

official field in the area of local concern, also, its jurisdiction is accepted with deference. The State, in all its domestic functions, is not a necessary evil as it is so generally felt to be in the United States. The State comes nearer, in Japanese eyes, to being the supreme good.

(81)

In this portrayal, Japanese society was depicted as hierarchically organized, with the state monopolizing significant power and effectiveness. While Benedict's anthropological work has faced criticism on various fronts, her core arguments still resonate in more contemporary works. For instance, Neher's analysis of Asian democracy asserts, "Asian democracy is characterized by respect for authority and hierarchy" (Neher 1994, 953). Asian societies continue to be characterized by "a universal moral order that remains in harmony as long as all individuals fulfill their duties by adhering to the obligations of subordinate to superior and vice versa" (Neher 1994, 954).

While the modernizing state may indeed hold considerable power in Japan, extending this view (which emphasizes universal respect for authority, hierarchy, harmony, and so on) to all aspects of society can be highly misleading. In this regard, the works of Kunio Yanagita (1875–1962), perhaps the most renowned scholar of folklore studies in Japan, are filled with specific examples that illustrate how various mechanisms moderated and even challenged social hierarchy in people's everyday lives. For example, regarding the issue of marriage in the prewar period, Benedict (1946) wrote,

In any affair of importance the head of a family of any standing calls a family council at which the matter is debated. For a conference on a betrothal, for instance, members of the family may come from distant parts of Japan. The process of coming to a decision involves all the imponderables of personality. A younger brother or a wife may sway the verdict. The master of the house saddles himself with great difficulties if he acts without regard for group opinion. Decisions, of course, may be desperately unwelcome to the individual whose fate is being settled. His elders, however, who have themselves submitted in their lifetimes to decisions of family councils, are impregnable in demanding of their juniors what they have bowed to in their day.

(53)

This illustration is meant to highlight the nature of reciprocal hierarchy, wherein the superior assumes both power and responsibility over subordinates. In contrast, Yanagita vividly portrays how gender-based social groups in a village played a key role in match-making:

When a couple are seriously in love, the man consults his group and the woman hers. If the prospective union is approved by their respective groups, the leaders of the *wakamono gumi* (young men's group) to go see the parents of both parties for approval of their marriage. If and when the parents do not agree to their marriage, the young men's group may put pressure upon the parents, They might threaten to refuse to help them repair their straw-thatch roof, or mend the road in front of their house, or extinguish fires, or call for a doctor when emergency arises. It is usually the parents who are forced to succumb to the will of the young people.... In this matter of marriage the egalitarian relationships of the peer groups are more significant than the hierarchical relationship of parents and children.

(Tsurumi 1975, 25)

Yanagita's insights call our attention to the co-presence of formal hierarchy along with substantive egalitarianism and a certain degree of freedom, and to the important role of local community-level organizations, rather than only individuals and households (as exemplified in Benedict's account), as critical units of agency.

Why were community-level organizations and institutions often overlooked? Yanagita's discussion suggests some answers to this question. He states that during the Meiji period in Japan, when the modern state replaced the feudal systems,

The former samurai, who became government officials, had long been prohibited from making voluntary associations among themselves. The only union they belonged to was their *han* (fief), which was large in size. That is why they lacked deep understanding of the history of the self-governing associations which had functioned among peasants, craftsmen, and merchants in the previous period. Thus the officials were led to discard many customs of the people which could have been adapted and profitably developed for use in the new era. Instead, the bureaucrats established various organizations and institutions without giving the people concerned any chance to discuss and study the adequacy of the new systems.

(Yanagita 1931, 345–346, translated by Tsurumi 1975, 30)

In other words, in the eyes of the modernizing elites, those "pre-modern" local social organization and institutions were either invisible or insignificant. It is possible that some observers of Japanese culture, such as Benedict, may have unconsciously treated the ethics, viewpoints, and cultures of the samurai class and the modernizing elites as representing the Japanese society. Another important point to note is that the "pre-modern" social institutions and organizations did not disappear immediately with the onset of modernization, although they may have undergone many changes. Some of these institutions may still exist today, or their underlying principles may have been inherited and continue to influence everyday life, particularly in rural areas, as many chapters in this book reveal.

Although Japan stands out as one of the few countries that escaped European colonization and underwent a unique modernization process, the dynamics it displayed between the modernizing state and the "pre-modern" spheres of everyday life for ordinary people may have relevance elsewhere. Kuan-Hsing Chen, a Taiwanese cultural theorist, provides an insightful discussion of the meaning and roles of "civil society" in Asia. Chen (2010) observes that the translation and understanding of "civil society" vary across East Asian countries. He argues that this linguistic fluidity signifies that concepts like civil society and citizenship "were not smoothly absorbed into the political culture of many spaces in Asia" during modernization (237).

In Asia, the term "civil society" is often associated with the modernizing state and with "moralizing words" such as "feudalistic," "antiquated," and "wasteful" used to describe and attack the practices and institutions of common people (Chen 2010, 240). However, Chen highlights the importance of the space where traditions are maintained as resources for common people to navigate the disruptive changes brought about by the modernizing state and civil society. In Chinese, this space is referred to as *minjian* (literally "people-space" or roughly translated as "folk society"). Concrete manifestations of *minjian* organizations can take various forms, such as local funeral groups, religious festival organizations, neighborhood associations, and mutual-aid or informal loan associations. Some observers in the West may perceive these as remnants of the past that can be disregarded in societal analysis and policymaking. Chen, however, argues that *minjian* "did not disappear during the process of modernization" (240) but rather evolved and operate visibly in the present (241). In many ways, the realm of *minjian* aligns with the focus of life-environmentalism, and the notion of *minjian* opens up space for productive engagement between life-environmentalism and a broader audience in other parts of Asia.

Intellectual Contexts

How Do Theories Travel (or Stay Home)?

In North American colleges and universities, concepts such as deep ecology and ecological modernization are commonly taught as ways of thinking about human–environment interactions. Similarly, environmental justice and political ecology have become influential analytical frameworks and fields of inquiry, to which entire college courses may be devoted. These concepts and approaches can be broadly categorized as "theories" and have gained significant visibility and global reach.

One definition of deep ecology is an "environmental philosophy and social movement based in the belief that humans must radically change their relationship to nature from one that values nature solely for its usefulness to human beings to one that recognizes that nature has an inherent value" (Madsen 2023). The origin of deep ecology is commonly attributed to a 1973 paper by

Arne Næss, a Norwegian philosopher, whose thinking was critically influenced by Rachel Carson's *Silent Spring* (1962). The concept of ecological modernization first emerged from the German environmental debate in the 1980s, and its theoretical development initially took place in selected western European countries, including Germany, the Netherlands, and the United Kingdom (Andersen and Massa 2000; Fisher and Freudenburg 2001; Mol and Sonnenfeld 2000). It suggests that "it is possible, through the development of new and integrated technologies, to reduce the consumption of raw materials, as well as the emissions of various pollutants, while at the same time creating innovative and competitive products" (Andersen and Massa 2000, 337). In short, deep ecology is typically associated with critical reflection on modern industrialization and valuing nature for its intrinsic worth, while ecological modernization envisions a kind of modernization that allows for economic development and environmental conservation to coexist (Bell 2011).

The concept of environmental justice has also gained significant influence as a perspective for analyzing and addressing socio-environmental issues. Its origins are usually traced back to the controversies surrounding the siting of hazardous waste landfills in predominantly African-American communities in the United States, notably Houston, Texas, in 1979 and Warren County, North Carolina, in 1982. One commonly used definition, particularly in the United States, is an emphasis on "the fair treatment and meaningful involvement of all people regardless of race, color, national origin, or income with respect to the development, implementation, and enforcement of environmental laws, regulations, and policies" (US EPA 2015). Over the years environmental justice approach has been applied to a tremendous range of locales (Holifield, Chakraborty, and Walker 2017), including Asia (Jobin, Ho, and Hsiao 2021).

Political ecology is not so much a single "theory" as it is a body of literature that encompasses a range of ideas and definitions. Nevertheless, the literature is grounded in a common premise that "environmental change and ecological conditions are the product of political process" and are best understood as such (Robbins 2012, 19–20). The influential works of political ecology have drawn from intensive fieldwork conducted in the Global South (e.g., Blaikie and Brookfield 1987), as well as investigations into environmental degradation in heavily industrialized countries (e.g., Cockburn and Ridgeway 1979). Today, it is arguably one of the most popular perspectives used to analyze various environmental issues within Anglophone social sciences.

All of these theories, described above, have gained recognition beyond their original contexts in the industrialized West. It is not surprising, as the core principles of these theories do have universal relevance. After all, who would outright reject the notion of the intrinsic value of "nature" (deep ecology) or the idea that economic growth can be compatible with environmental sustainability (ecological modernization)? Likewise, the reality of marginalized populations disproportionately bearing environmental burdens, a central tenet of environmental justice and political ecology, can undoubtedly be observed around the world.

The widespread dissemination of concepts and theories had a positive outcome: they undergo modifications, expansions, and further refinements. For instance, as environmental justice activism and research have broadened in terms of themes and geographical locations, the meanings and dimensions of environmental justice and related ideas have evolved and diversified. The groups seen to be affected by environmental injustice have expanded beyond racialized minority and low-income populations (which were the original focus of research in the United States) to include indigenous populations and even nonhuman entities (Holifield, Chakraborty, and Walker 2017). In Asia, as noted by Jobin, Ho, and Hsiao (2021), environmental justice is frequently invoked to address class struggles and land conflicts between the dominant group and ethnic minorities or indigenous peoples, rather than focusing primarily on racism as understood in the U.S. context with explicit references to skin color.

Moreover, new dimensions of environmental justice have been articulated over time. In addition to distributive justice, studies now emphasize procedural and participatory justice, justice as recognition, and justice as capabilities. In this way, the literature on environmental justice has established itself as a significant lens through which to understand contemporary socio-environmental issues, appealing to a wide range of activists, scholars, and practitioners worldwide. In short, as the literature on environmental justice continues to spread, it becomes more diverse, its theoretical insights gain broader applicability, and it provides valuable policy implications.

While the dissemination of theories is generally welcomed and even celebrated, a significant issue arises from the current one-sided nature of this dissemination. The concepts and theories mentioned above have mostly been developed or articulated by scholars in Euro-American academic settings and have gained popularity in other regions. However, the reverse flow, where theories and concepts from non-Western contexts gain traction in Euro-American scholarship, is much weaker. Figures 1.5 and 1.6 compare popular concepts and theories in contemporary Anglophone and Japanese scholarship. In both scholarly communities, approaches of Western origin such as "environmental justice" and "political ecology" are prominently featured. However, despite its notable presence within Japan, life-environmentalism remains largely unknown in the Anglophone literature. The majority of scholarship informed by life-environmentalism has been published in Japanese, with a few noteworthy exceptions (Kada 2006; Furukawa 2007; Torigoe 2014; Noda 2017; Kaneko 2017; Ueda and Torigoe 2012). Even the available English-language works on the topic do not extensively engage with contemporary Anglophone scholarship.

Japanese academics sometimes joke that the Japanese academia is an intellectual black hole—vast amounts of ideas and theories from outside (mostly from the West) are continuously absorbed, but little emerges in return. While this statement may be an exaggeration, it reflects a sentiment and situation that may resonate with scholars primarily working outside the Anglophone academic context. Factors that contribute to this situation include language barriers,

Introduction 13

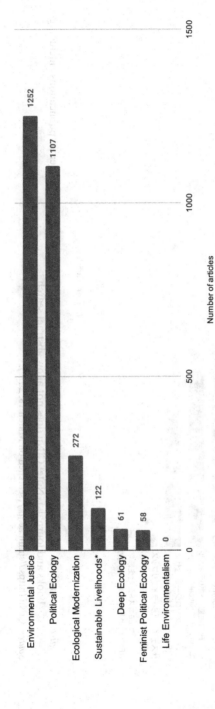

Figure 1.5 Number of articles with the term appearing in the abstract of peer-reviewed articles in English-language scholarly journals. Note: "Sustainable Livelihood" includes both "sustainable livelihood(s) approach" and "sustainable livelihood(s) framework."

Source: International Bibliography of Social Sciences (IBSS) (searched on April 28, 2020).

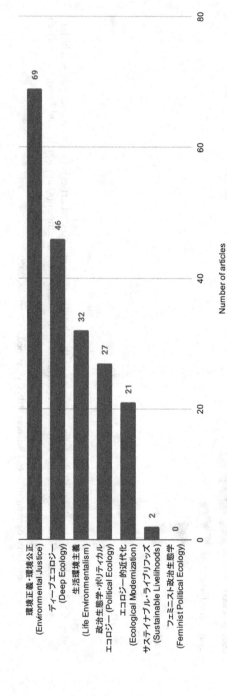

Figure 1.6 Numbers of articles with the term appearing in the title of Japanese-language academic publications (including books and journals). Note: CiNii does not allow abstract search.

Source: CiNii (https://cir.nii.ac.jp/) (searched on March 18, 2022).

limited human resources and funding, academic politics, and a lack of mutual interest. Nonetheless, considering the potential costs of such asymmetric flows of ideas, it is reasonable to express concern and strive for greater interchange.

Imposition of Infeasible or Problematic Solutions

As concepts and theories travel and find application in various real-world contexts, observers and practitioners may consciously or unconsciously demand "drastic solutions" or advocate for "questionable" changes. These demands reveal critical assumptions that underlie the concepts and theories themselves. For instance, in the 1980s, when deep ecologist Roderick Nash told Indian historian Ramachandra Guha, "the less developed nations may eventually evolve economically and intellectually to the point where nature preservation is more than a business" (quoted in Guha 2003, 27), the implicit message was that environmental conservation necessiate "modernization" as a prerequisite. From Guha's perspective, deep ecology was primarily an environmentalism of the affluent world, focusing on wilderness preservation while disregarding rampant material consumption. At its extreme, this perspective viewed "nature appreciation as an indication of cultural maturity" (Guha 1990, 435).[3]

Ecological modernization, for its part, places difficult demands on non-Western countries. Firstly, ecological modernization emerged in Europe during a historical context where the transition toward a service-based economy and the decline of polluting manufacturing industries were already taking place. However, these polluting industries were being relocated to countries in Asia and Africa (Bell 2011, 33). Advocating for ecological modernization in non-Western countries implicitly demands "modernization" while accepting the presence of "polluting industries" or their relocation elsewhere.

Secondly, ecological modernization assumes and promotes the idea of "civil society" as an essential component for its implementation. However, civil society is often perceived as underdeveloped or underperforming in non-Western contexts. For example, Mol (2006, 52), a prominent advocate of ecological modernization theory, argues that "civil society in China remains undeveloped and has been unable to match the role played by civil society institutions and actors in most OECD countries." While acknowledging the development of unique nonstate, nonmarket actors like "Government-organized NGOs (GONGOs)" in China, he considers a stronger role for civil society crucial for promoting ecological modernization. Its achievement requires greater access to environmental information, which in turn necessitates overcoming various obstacles such as the lack of environmental monitoring, information distortion, secrecy surrounding environmental data, limited right-to-know legislation or practice, and restricted publication and availability of nonsecret data due to poor reporting and limited internet use and access (Mol 2006, 49). Recent examples, such as the Chinese government's control over internet use and access, demonstrate the challenges in achieving greater transparency and access to environmental information.

Regarding environmental justice, Williams and Doyon (2019) point out that struggles for environmental justice in postcolonial India have been dominated by the educated and well-connected middle class (668). They also note that much environmental justice research relies on "the specific historical experience of the West, where democratic systems, for all their faults, have achieved a degree of formal equality of recognition" (668). In certain parts of Asia, where the concept of environmental justice has gained significant traction and popular support, such as Taiwan, environmental movements and democratization have been inseparable processes (Fan and Chou 2017). This means that demanding environmental justice involves pursuing not only justice in environmental matters but also substantive democracy. However, this can be a hard task in authoritarian regimes. Jobin, Ho, and Hsiao (2021) assert that environmental activism in the context of authoritarian regimes in Asia is a "dangerous endeavor," putting local actors at risk of intimidation, imprisonment, torture, murder, rape, and other human rights violations. Even in less-extreme situations, imposing formulaic democratic frameworks and procedures may create divisions and distress within societies without effectively addressing the problem at hand.

In short, it is important to consider the broader economic, political, and societal conditions from which these concepts and theories originate or that they assume. These conditions include functioning democratic institutions, robust civil society, a history of civil rights movements, recognition of racism as a societal issue, and a certain level of industrialization and technological development already achieved. Consequently, when a "problem" is identified in non-Western contexts, observers may focus on the perceived lack, deficiencies, dysfunction, and immaturity of what the West already possesses (such as dysfunctional democratic and judicial systems, weak civil society, lack of civil rights movements, nonrecognition of racism, or limited industrialization). The non-West is often seen as the "deficit other," and outside observers may implicitly or explicitly endorse or demand wholesale institutional or societal transformations to solve environmental problems. However, it is important to recognize that local communities may not necessarily view a comprehensive sociopolitical transformation as the sole or most promising solution to their environmental issues.

Impact on Theoretical Development

Another consequence of the dominance of Euro-American concepts and theories is their impact on theoretical development. As Jobin, Ho, and Hsiao (Jobin, Ho, and Hsiao 2021) aptly point out, referring to theoretical bodies of literature like environmental justice and political opportunity theory, "this intellectual domination of North American literature tends to neglect other theoretical perspectives and minimize cultural specificities when, at a critical time for biodiversity, we should also pay attention to a broader diversity of theoretical approaches" (13). This dominance can affect theoretical development in several ways.

One way is that issues that fall outside the scope of dominant theories tend to receive less attention or remain less visible. For example, in research on environmental justice, Fan and Chou (2017) observe that "literature on environmental justice in Japan is scant and recent" (620). Despite the existence of many studies of events like the Ashio Copper Mine Incident (1890s) and major pollution episodes in the 1950s–1970s, such as Minamata mercury disease and Yokkaichi asthma, which sparked large antipollution movements and highlighted the disproportionate environmental burden faced by marginalized populations, these studies are not commonly considered works on environmental justice. This invisibility is likely because they do not explicitly use the language of "environmental justice" and because there is relatively limited reference in the English-language literature to these topics (for English-language studies, see Walker (2010), George (2001), and Funabashi (2006)).

Another related impact is that the dominance of particular theories may lead to a focus on specific empirical issues that align with those theories. It is undeniable, for instance, that "in many countries, including in Asia, environmental activists mobilized under a rally call for EJ (Environmental Justice)" (Jobin, Ho, and Hsiao 2021, 13). When there are clear perpetrators on one side and unquestionable victims on the other, environmental justice can serve as a suitable and powerful analytical tool. However, there are other environmental issues that affect ordinary residents of local communities but do not display clear divisions of perpetrators and sufferers. The way environmental activists frame the issue may differ significantly from how local residents perceive it. In short, there is a risk of falling into the trap of "when you have a hammer, everything looks like a nail"—students who are aspired to find environmental injustice may be tempted to see only that in the field.

Yet another effect is that local conditions, resources, and assets that do not fit into Western categories appear less visible, less central, less complete, less useful, and less contemporary. As Chatterjee (1997) states, "the normative models of Western political theory have, more often than not, only served to show non-Western practices as backward or deviant" (282). For example, in the study of Kaminoseki Town, where a proposed nuclear power plant has sparked a lasting controversy, Dusinberre (2012) explains that local civil society actors, such as fishing/farming cooperatives, the chamber of commerce, and PTAs, operate within a "democratic deficit" where "many of the so-called horizontal associations within civil society were only nominally autonomous from local government and thus, by extension, from the state itself" (Dusinberre 2012). In this perspective, Japanese "civil society" is seen as an incomplete version of what it should be. To be fair, this is not solely the fault of external observers. It is also because "we [Asians] are constantly being pushed by our nationalist elites to follow and catch up with the West by reproducing those Euro-American forms of civil society" (Chen 2010, 243). This aligns with the point Chen makes when discussing *minjian* as an active social space whose power "still has to be dealt with carefully, even by the authoritarian state" (239). Examples of the role of *minjian* include events and activities that revolve around the lunar calendar, "informal" economic activities, mutual-aid groups, and more. Instead of viewing them as fading cultural legacies, can we envision theories in which these "pre-modern" institutions and practices also play pivotal roles alongside modernizing institutions?

Finally, the uncritical embrace of Euro-American theories not only makes on-the-ground problems and potential resources/assets less visible or more deviant, but it also delegitimizes other forms of knowledge. This is a central critique put forward by feminist and indigenous studies scholarship (e.g., Gibson-Graham 2006; Smith 2012). For instance, Coombes et al. (2012), in their review of indigenous geographies, recognize some shared perspectives between political ecology—currently a popular perspective in Anglophone cultural anthropology and human geography—and indigenous geographies, such as a critical stance toward the "off the shelf," modern environmental management imposed by the state on indigenous peoples. However, they express concern about political ecologists' persistent disregard for culture, their ethnocentric assumptions about indigenous practices, and their apparent resistance to accepting ontological pluralism (812–813). Accordingly, Coombes et al. (2012) argue that:

> Political ecologists' critique of community sometimes suggests that Indigenous claims as to its importance be treated as false consciousness. Watts (2006: 103) assessed the political use of chieftainship, indigeneity and community in Nigeria, concluding that they are "imagined" and violent. Communitarian discourses may protect economic interests and that demands analysis, but Watts' universalizing disrespect for those Indigenes who organize themselves around the concept of community is ethnocentric.
> (812)

Watts' view certainly does not represent the entire field of political ecology, but it does illustrate the difficulties in incorporating what indigenous peoples may consider central and real in their lives into dominant theories. Furthermore, Coombes et al. (2012) allude to a profound ontological difference that may not be easily reconciled.

> Although political ecologists claim unique perception of nature's agency [labeling their focus to be "more-than-human geographies"], it is unclear whether they will accede to the possibility of ontological difference and multinaturalism.
> (813)

In this context, multinaturalism (rather than multiculturalism) refers to the recognition that indigenous peoples and colonizing others possess fundamentally different worldviews (812). Although the Japanese, particularly the Yamato people, may not typically be considered "Indigenous," the observation about multinaturalism is still relevant in this context.

Life-environmentalism as a Weak Theory

Murton (2012) argues for the need to demonstrate how indigenous philosophies "can work with European philosophy to make a difference both politically and intellectually" (88). In this regard, Murton suggests that Maori knowledge systems, which emphasize genealogy as a key element of knowledge, a non-individualistic understanding of the self, and the foundational role of speech and naming in the acquisition of knowledge, can be connected to European philosophical traditions represented by thinkers like Martin Heidegger and Maurice Merleau-Ponty.

While establishing such connections may be a valuable endeavor, it should not be a prerequisite for recognizing the validity of indigenous philosophies. Additionally, presenting non-Western intellectual traditions within the framework of Western traditions and using their language runs the risk of merely reinforcing the apparent versatility and diversity of Western scholarship. This book itself carries that risk. Life-environmentalism may not seem particularly groundbreaking to those familiar with contemporary Euro-American scholarship on socio-environmental issues and sustainability. The language of life-environmentalism may appear somewhat eccentric, but some may suggest that its fundamental principles and many empirical insights can be effectively conveyed using existing analytical frameworks such as political ecology. In response, I draw upon the spirit of "weak theory" as discussed by Gibson-Graham (2008), who state:

> What if we were to accept that the goal of theory is not to extend knowledge by confirming what we already know...? What if we asked theory instead to help us see openings, to provide a space of freedom and possibility?
>
> (619)

According to Gibson-Graham (2008), the "strong theory" reinforces our sense of powerlessness by asserting the absolute and exploitative logic of neoliberal capitalist globalization. They argue instead for "the practice of weak theorizing [which] involves refusing to extend explanation too widely or deeply, refusing to know too much" (619). I interpret the value of weak theory to hinge on whether theory is "actionable" or not, which means whether it resonates with the minds and bodies of those who are subjected to analysis or being theorized.

Torigoe (2002) states that life-environmentalism is strongly influenced by the intellectual tradition of Japanese folklore studies, particularly the work of Kunio Yanagita. This tradition regards *kokoro* (often translated as "mind" or "heart") as an essential quality and subject of intellectual inquiry.[4] Torigoe (2002, iii) asserts,

> Euro-American social science lacks a methodological framework for the concept of *kokoro*. While the Freudian approach may be intellectually comprehensible and intriguing to us, it does not resonate with our body.

> When it comes to social issues ... it would be nice and neat if we interpret environmental problems using an ecological approach, which our head can easily understand. However, there is something that our living bodies do not fully approve of.
>
> (author's translation)

If life-environmentalism holds any value, it is because it resonates more effectively with those whom the theory aims to understand, rather than because of its general applicability or its connection to other theoretical traditions. I often encounter students who say, "life-environmentalism is interesting, but it does not seem to work here (i.e., in the settler society of the United States)." In a sense, that's precisely the essence of life-environmentalism.

Nevertheless, there have been various important criticisms directed at life-environmentalism. Some of these criticisms and debates are addressed in the chapters included in this book, and further engagement with diverse theoretical traditions and empirical studies from different parts of the world is clearly welcome.

The Rest of the Book

In this chapter, I have tried to go beyond the scope of life-environmentalism's original proponents to position it in broader contexts. The next chapter (Chapter 2) provides a conceptual overview of life-environmentalism, highlighting key epistemological characteristics of the approach. Chapters 3–14 are empirical case studies, all of which are informed by life-environmentalism, or more broadly the *seikatsu-ron* school of thought, to varying degrees. I believe that the case studies are where life-environmentalism research truly excels, and that its empirical insights are what its value should ultimately be judged against.

Chapters 3–5 focus on how local communities respond to various external shocks, broadly described as "developmental shocks," including the pressure to promote tourism, build leisure facilities, and consent to infrastructure projects. Each chapter sheds light on how different kinds of everyday-life organizations, locally specific rules, and conventions mitigate the "shock" emanating from developmental shocks. Chapters 6–8 analyze how local communities govern and manage their everyday space under formidable circumstances, often skillfully navigating around powerful actors such as private corporations and state agents. Notably, none of these chapters is about residents overtly seeking "justice" or is about modern civil society organizations.

Chapters 9–11 examine local communities in the Tōhoku region that were affected by the earthquake–tsunami–nuclear disaster in 2011. Each of these chapters reveals locally specific ways in which inhabitants have dealt with disasters such as tsunamis and famines, the loss of their families and relatives, and the challenges of material and spiritual recovery. Chapters 12–14 deal with what may be called human interactions with historically-formed socio-cultural environment in urbanized settings. While life-environmentalist work often focuses on rural areas and the "natural" environment, these chapters point to the relevance of the life-environmentalist approach in certain non-rural settings. Chapters 15–17 offer critical reflections and discussion of key debates and prospects of the life-environmentalism research. Finally, Chapter 18 offers the translation of selected texts from one of the pioneering books on life-environmentalism (Torigoe 1989) as key reference for the readers.

It is our hope that this book offers an opportunity to imagine alternative ways of thinking about socio-environmental challenges and prospects for creative intellectual engagements. Life-environmentalism developed a relatively coherent way of thinking about socio-environmental problems in late-20th-century Japan. It uses words and ideas, such as *kokoro*, *seikatsu*, and *iibun*, that resonate more readily with the ordinary people it tries to describe and understand. At the same time, in my view, the language of life-environmentalism is "logical" enough to facilitate productive conversations with contemporary English-language social scientific literature on human–environment relationships and sustainable development.

Notes

1 In previously published English writings, *Seikatsu Kankyō Shugi* has been translated as "life environmentalism" without a hyphenation (e.g., Kada 2006; Torigoe 2014; Yamamoto 2019). See Chapter 2 for the rationale behind our use of "life-environmentalism" in this book.
2 There are also significant debates about the nature, evolution, and variety of "democracy" in Asian countries. For example, Iwazaki (1997) asserts, "The necessary perspective when analyzing the democracy of Asian countries is, rather than asking to what extent Asian countries have 'democratized' themselves, using Euro-American countries as the model, to ask how political elites of Asian countries understood democracy" (9).
3 It is important to note that Guha does not reject all aspects of deep ecology. In his proposal for a new environmental philosophy, called ecological socialism, he embraces the core idea of *diversity* from the deep ecological ethic.
4 Lafcadio Hearn, also known as Yakumo Koizumi in Japan, described the term *kokoro* as encompassing "…mind, in the emotional sense; spirit; courage; resolve; sentiment; affection; and inner meaning, – much like the way we express it in English as 'the heart of things'" (Hearn 1907, 11).

References

Andersen, M. S., and I. Massa. 2000. "Ecological Modernization—Origins, Dilemmas and Future Directions." *Journal of Environmental Policy and Planning* 2: 337–45.
Bell, Sarah. 2011. *Engineers, Society, and Sustainability*. Cham: Springer.
Benedict, Ruth. 1946. *The Chrysanthemum and the Sword: Patterns of Japanese Culture*. Boston: Houghton Mifflin.
Blaikie, Piers M., and H. C. Brookfield. 1987. *Land Degradation and Society*. London, New York: Methuen.
Bolt, Jutta, and Jan Luiten van Zanden. 2020. Maddison Project Database, Version 2020. https://www.rug.nl/ggdc/historicaldevelopment/maddison/releases/maddison-project-database-2020

Castells, Manuel. 1992. "Four Asian Tigers with a Dragon Head: A Comparative Analysis of the State, Economy and Society in the Asian Pacific Rim." In *States and Development in the Asia Pacific Rim*, 33–70, edited by Richard P. Appelbaum and J. W. Henderson. Newbury Park, CA: Sage Publications.

Chatterjee, Partha. 1997. *A Possible India: Essays in Political Criticism*. Delhi and New York: Oxford University Press.

Chen, Kuan-Hsing. 2010. *Asia as Method. Toward Deimperialization*. Durham, NC: Duke University Press.

Cockburn, Alexander, and James Ridgeway. 1979. *Political Ecology*. New York: Times Books.

Coombes, Brad, Jay T. Johnson, and Richard Howitt. 2012. "Indigenous Geographies I: Mere Resource Conflicts? The Complexities in Indigenous Land and Environmental Claims." *Progress in Human Geography* 36 (6): 810–21. https://doi.org/10.1177/0309132511431410

———. 2013. "Indigenous Geographies II: The Aspirational Spaces in Postcolonial Politics – Reconciliation, Belonging and Social Provision." *Progress in Human Geography* 37 (5): 691–700. https://doi.org/10.1177/0309132512469590

———. 2014. "Indigenous Geographies III: Methodological Innovation and the Unsettling of Participatory Research." *Progress in Human Geography* 38 (6): 845–54. https://doi.org/10.1177/0309132513514723

Deans, Phil. 1999. "The Capitalist Developmental State in East Asia." In *State Strategies in the Global Political Economy*, edited by Ronen Palan, Jason Abbott, and Phil Deans. 78–102. London: Continuum International Publishing Group.

Duara, Prasenjit. 2014. *The Crisis of Global Modernity. Asian Traditions and a Sustainable Future*. Cambridge: Cambridge University Press.

Dusinberre, Martin. 2012. "DIMBY: Kaminoseki and the Making/Breaking of Modern Japan." *The Asia–Pacific Journal: Japan Focus* 10 (32). https://doi.org/10.5167/uzh-116315

Ethnologue. 2023. Languages of the World. https://www.ethnologue.com/

Fan, Mei-Fang, and Kuei-Tien Chou. 2017. "Environmental Justice in a Transitional and Transboundary Context in East Asia." In *The Routledge Handbook of Environmental Justice*, 615–26, edited by Ryan Holifield, Jayajit Chakraborty, and Brian Walker. London: Routledge.

Fisher, D., and W. Freudenburg. 2001. "Ecological Modernization and Its Critics: Assessing the Past and Looking toward the Future." *Society and Natural Resources* 14: 701–9.

Fisher, Max. 2015. "Map: European Colonialism Conquered Every Country in the World but These Five." *Vox*. Accessed July 9, 2023. https://www.vox.com/2014/6/24/5835320/map-in-the-whole-world-only-these-five-countries-escaped-european

Funabashi, H. 2006. "Minamata Disease and Environmental Governance." *International Journal of Japanese Sociology* 15 (1): 7–25.

Furukawa, Akira. 2007. *Village Life in Modern Japan: An Environmental Perspective*. Melbourne: Trans Pacific Press.

George, Timothy S. 2001. *Minamata: Pollution and the Struggle for Democracy in Postwar Japan*. Cambridge, MA: Harvard University Press.

Gibson-Graham, J. K. 2006. *A Postcapitalist Politics*. Minneapolis, MN: University of Minnesota Press.

———. 2008. "Diverse Economies: Performative Practices for 'Other Worlds'." *Progress in Human Geography* 32 (5): 613–32.

Goldewijk, Klein. 2022. History Database of the Global Environment 3.1. https://public.yoda.uu.nl/geo/UU01/G4HO5I.html

Guha, Ramachandra. 1990. "Toward a Cross-Cultural Environmental Ethic." *Alternatives: Global, Local, Political* 15 (4): 431–47.
———. 2003. "How Much Should a Person Consume?" *Vikalpa* 28 (2): 1–11.
Hearn, Lafcadio. 1907. *Kokoro: Hints and Echoes of Japanese Inner Life*. Collection of British Authors. Tauchnitz Ed., 3957. Leipzig: Bernhard Tauchnitz.
Holifield, Ryan, Jayajit Chakraborty, and Gordon Walker. 2017. *The Routledge Handbook of Environmental Justice*, 1st ed. London: Routledge. https://doi.org/10.4324/9781315678986
Iwazaki, Ikuo, ed. 1997. *Ajia to Minshu Shugi: Seiji Kenryokusha no Shisō to Kōdō* [Asia and Democracy: Thoughts and Actions of Political Elites]. Tokyo: IDE-JETRO.
Jobin, Paul, Ming-sho Ho, and Hsin-Huang Michael Hsiao. 2021. "Environmental Movements and Politics of the Asian Anthropocene: An Introduction." In *Environmental Movements and Politics of the Asian Anthropocene*, 1–36, edited by Paul Jobin, Ming-sho Ho, and Hsin-Huang Michael Hsiao. ISEAS–Yusof Ishak Institute.
Johnson, Chalmers. 1999. "The Developmental State: Odyssey of a Concept." In *The Developmental State*, 32–60, edited by Meredith Woo-Cumings. Ithaca, NY: Cornell University Press.
Kada, Yukiko. 2006. "Three Paradigms behind River Governance in Japan: Modern Technicism, Nature Conservationism and Life Environmentalism." *International Journal of Japanese Sociology* 15 (1): 40–54.
Kaneko, Hiroyuki. 2017. "Radioactive Contamination of Forest Commons." In *Unravelling the Fukushima Disaster*, 136–53, edited by Mitsuo Yamakawa and Daisaku Yamamoto. London: Routledge.
Kawanabe, Hiroya, Machiko Nishino, and Masayoshi Maehata. 2012. *Lake Biwa: Interactions between Nature and People*. Dordrecht: Springer. http://ebookcentral.proquest.com/lib/colgate/detail.action?docID=971402
Madsen, Peter. 2023. "Deep Ecology." *Encyclopedia Britannica*. https://www.britannica.com/topic/deep-ecology
Mol, Arthur P. J. 2006. "Environment and Modernity in Transitional China: Frontiers of Ecological Modernization." *Development and Change* 37 (1): 29–56.
Mol, Arthur P. J., and David A. Sonnenfeld. 2000. "Ecological Modernisation around the World: An Introduction." *Environmental Politics* 9 (1): 1–14. https://doi.org/10.1080/09644010008414510
Muldavin, Joshua S. S. 2008. "The Politics of Transition: Critical Political Ecology, Classical Economics, and Ecological Modernization Theory in China." In *The SAGE Handbook of Political Geography*, edited by Kevin R Cox, Murray Low, and Jennifer Robinson, 247–62. Los Angeles: SAGE Publications.
Murton, Brian. 2012. "Being in the Place World: Toward a Māori 'Geographical Self'." *Journal of Cultural Geography* 29 (1): 87–104.
Neher, Clark D. 1994. "Asian Style Democracy." *Asian Survey* 34 (11): 949–61.
Noda, Takehito. 2017. "Why Do Local Residents Continue to Use Potentially Contaminated Stream Water after the Nuclear Accident?" In *Rebuilding Fukushima*, 53–68, edited by Mitsuo Yamakawa and Daisaku Yamamoto. London: Routledge.
Our World in Data. 2023. Political Regime, 2019. https://ourworldindata.org/grapher/political-regime?time=2019
Pakrashi, D., and P. Frijters. 2017. *Takeoffs, Landing, and Economic Growth*. ADBI Working Paper 641. Tokyo: Asian Development Bank Institute. https://www.adb.org/publications/takeoffs-landing-and-economic-growth
Robbins, Paul. 2012. *Political Ecology: A Critical Introduction*, 2nd ed. Chichester, West Sussex: John Wiley & Sons.

Rudolph, Lloyd I., and Susanne Hoeber Rudolph. 1987. *In Pursuit of Lakshmi: The Political Economy of the Indian State*. Chicago: University of Chicago Press.
Smith, Linda Tuhiwai. 2012. *Decolonizing Methodologies: Research and Indigenous Peoples*, 2nd ed. London: Zed Books.
Torigoe, Hiroyuki, ed. 1989. *Kankyō Mondai no Shakai Riron: Seikatsu Kankyō Shugi no Tachiba Kara* [Social Theory of Environmental Problems: From the Standpoint of Life-environmentalism]. Tokyo: Ochanomizu Shobō.
Torigoe, Hiroyuki. 2002. *Yanagita Minzokugaku no Firosofi* [*Yanagida Kunio and Japanese Folklore*]. Tokyo: University of Tokyo Press.
———. 2014. "Life Environmentalism: A Model Developed under Environmental Degradation." *International Journal of Japanese Sociology*, 23 (1): 21–31.
Torigoe, Hiroyuki, and Yukiko Kada, eds. 1984. *Mizu to Hito no Kankyōshi: Biwako Hōkoku Sho* [Environmental History of Water and People: Lake Biwa Report]. Tokyo: Ochanomizu Shobō.
Tsurumi, Kazuko. 1975. *Yanagida Kunio's Work as a Model of Endogenous Development*. Tokyo: Institute of International Relations, Sophia University.
Ueda, Kyoko, and Hiroyuki Torigoe. 2012. "Why Do Victims of the Tsunami Return to the Coast?" *International Journal of Japanese Sociology*, 21 (1): 21–9.
US EPA. 2015. "Learn About Environmental Justice." https://www.epa.gov/environmentaljustice/learn-about-environmental-justice
Walker, Brett L. 2010. *Toxic Archipelago: A History of Industrial Disease in Japan*. Seattle: University of Washington Press. https://www.jstor.org/stable/j.ctvcwn303
Watts, Michael. 2006. "The Sinister Political Life of Community: Economies of Violence and Governable Spaces in the Niger Delta, Nigeria." In *The Seductions of Community: Emancipations, Oppressions, Quandaries*, 101–42, edited by Gerald W Creed. Santa Fe: School of American Research Press.
Whittaker, D. Hugh, Timothy Sturgeon, Toshie Okita, and Tianbiao Zhu. 2020. *Compressed Development: Time and Timing in Economic and Social Development*. Oxford, UK: Oxford University Press.
Williams, Glyn, and Emma Mawdsley. 2006. "Postcolonial Environmental Justice: Government and Governance in India." *Geoforum* 37 (5): 660–70. https://doi.org/10.1016/j.geoforum.2005.08.003
Williams, Stephen, and Andréanne Doyon. 2019. "Justice in Energy Transitions." *Environmental Innovation and Societal Transitions* 31 (June): 144–53. https://doi.org/10.1016/j.eist.2018.12.001
Woo-Cumings, Meredith. 1999. *The Developmental State*. Ithaca, NY: Cornell University Press.
Yamamoto, Daisaku. 2019. "Life Environmentalism." In *Wiley Blackwell Encyclopedia of Sociology*, edited by George Ritzer. Oxford, UK: John Wiley & Sons.
Yanagita, Kunio. 1931. *Meiji Taishō Shi Dai 4 Kan Sesōhen* [History of Meiji and Taishō Vol. 4: Social Conditions]. Tokyo: Asahi Shimbun Sha.
Yeh, Emily T. 2009. "Greening Western China: A Critical View." *Geoforum; Journal of Physical, Human, and Regional Geosciences* 40 (5): 884–94. https://doi.org/10.1016/j.geoforum.2009.06.004

2 Theorizing Everyday Life
The Life-Environmentalist Way

Daisaku Yamamoto

Basic Propositions

In Chapter 1, life-environmentalism is described as a theory in a broad sense. However, it has been referred to in various ways, including as a "position/ standpoint (*tachiba*)" (Torigoe 1989; Torigoe and Kada 1984), a "model" (Torigoe 2014, 21), an "approach" and "method" (Furukawa 2007), and a "paradigm" for "an ideal type of environmental policy" (Kada 2006). This raises a question about the nature of the term itself.

In Japanese "life-environmentalism" is written in six Chinese characters as a compound of three two-character words: *seikatsu* (生活: life/livelihood/living), *kankyō* (環境: environment), and *shugi* (主義: ideology/-ism).[1] The use of the term *shugi* (-ism) is intentional; the creators of the term believed that it was important to emphasize that environmental research could not be free from ideological positions (Torigoe 1989, 20). When life-environmentalism was first proposed in the 1980s, the advocates of life-environmentalism were occasionally criticized by their fellow social scientists for not being "scientific," which supposedly meant being objective and impartial. Therefore, setting aside the question of whether life-environmentalism is a model, theory, or approach, it is worth noting that life-environmentalism is an ideologically self-aware proposition implying a particular political commitment, championing the concept of "life-environment."

But what is exactly the intellectual proposition that champions "life-environment?" Rather than explicitly defining what "life-environment" is, original proponents of life-environmentalism articulated their stance by comparing it with two other ideological positions: natural environmentalism (*shizen kankyō shugi*), which values the preservation of supposedly pristine nature (e.g., protecting or restoring a natural river), and modern technocentrism (*kindai gijutsu shugi*), which relies on modern technologies to overcome environmental problems (e.g., civil engineering projects for flood prevention). The former is akin to deep ecology, and the latter to ecological modernization in contemporary Euro-American environmental studies.

Life-environmentalists argued that the main difference between the idealized positions of natural environmentalism and modern technocentrism lay in the degree of human domination over nature (horizontal axis in Figure 2.1). In the 1970s–1980s, Japanese policy debates on environmental governance often revolved around this axis, focusing on whether to preserve nature or develop infrastructure. As briefly mentioned in Chapter 1, extensive embankment projects were initiated along the shores and streams of the Lake Biwa region in the 1970s. Local governments actively encouraged communities to "improve" these streams by reinforcing their beds and banks with concrete. The aim was to control flood risks and eliminate the need for regular, labor-intensive stream maintenance. Concrete would hinder plant growth, and any obstacles would be easily washed away. Unsurprisingly, such construction projects were vehemently opposed by (natural) environmentalists as they would destroy the local aquatic environment.

The team of researchers, who would later advocate for life-environmentalism, supported an alternative solution based on the insights they gained from their fieldwork. One significant discovery was that the stream water in communities that completely filled their streams with concrete gradually became dirtier compared to communities that maintained sand stream beds. This was because residents felt little incentive to regularly maintain the streams when they had concrete walls, and they saw less value in them. The streams became increasingly disconnected from residents' everyday lives and served merely as drains for household wastewater. Building on these insights, life-environmentalists proposed enclosing both banks of the stream with a permeable retaining wall while leaving the streambed as natural sand. This approach would allow aquatic

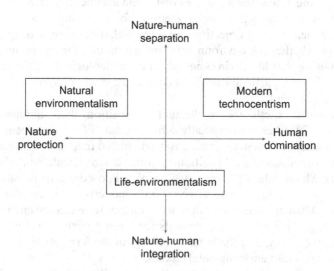

Figure 2.1 Positioning of life-environmentalism. Axes imply desirable nature–human relations. Modified based on Kada (2006).

species to thrive while providing some level of flood control (Yamamoto 2019; Torigoe 2014 provide a more comprehensive illustration in English).

At first glance, the solution endorsed by life-environmentalism may appear as a mere compromise between natural environmentalism and modern technocentrism. However, as explained by Kada (2006), life-environmentalism actually presents an alternative axis for understanding the human–environment relationship, which focuses on the degree of integration between nature and humans (vertical axis in Figure 2.1). While both natural environmentalism and modern technocentrism (intentionally or unintentionally) support the functional separation of everyday human activities from nature, life-environmentalism advocates for close functional proximity and integration between the two. Seen in this light, proponents of life-environmentalism in the case of the Lake Biwa embankment had a more compelling reason for endorsing the solution that appeared to be a mere compromise. The approach that they supported allows, encourages, and even requires residents to actively and regularly use and maintain the stream. This includes activities such as washing food, extinguishing fires, playing, fishing, and periodically clearing water weeds. This illustration shows that life-environmentalism recognizes that environmental conservation is best achieved by protecting the everyday life activities of a community that actively utilizes and manages its environmental resources.

These assertions of life-environmentalism are likely specific to the historical and geographical context of the communities in Japan where they are situated (Chapter 1). Some Euro-American audiences may argue, for example, that "industrial farmers also use natural resources, such as groundwater, on a regular basis, but that has clearly not resulted in better conservation." This highlights the need for caution in extending the claims of life-environmentalism too broadly and without careful consideration. Instead, it provides us with opportunities to define and specify the conditions under which life-environmentalism offers useful analysis and those under which it does not.

The above articulation of life-environmentalism's fundamental stance, which involves endorsing the integration of humans and nature by contrasting it with other approaches, provides a convenient way to introduce the concept. However, if that is the entirety of it, then there is little novelty in life-environmentalism. Many people would agree that the "integration of humans and nature" is a desirable principle. Therefore, let us delve deeper into the epistemological characteristics of life-environmentalism.

Life Beyond Survival

To start, the Japanese word "*seikatsu*" (*shenghuo* in Chinese and *saeng-hwal* in Korean) primarily refers to a way or manner of everyday living, or lifeways. This encompasses various aspects such as what and how people eat, dress, find shelter, interpret the world (including religious and belief systems), raise children, care for others, work and earn a living, maintain good health, and so on. Importantly, the word carries the nuance of concrete "everydayness."[2]

The nuance of *seikatsu* can be also compared with that of *seizon* ("survival"). The former signifies the concreteness and everyday nature of human activities that give meaning and uniqueness to their lives, while the latter focuses on the biological sustenance of life. Torigoe (2004) reinforces the significance of the notion of "life" by citing a comment from a speaker during the public hearings of the World Commission on Environment and Development in Sao Paulo in 1985:

> You [authors' note: panellists of the session] talk very little about life, you talk too much about survival. It is very important to remember that when the possibilities for life are over, the possibilities for survival start. And there are peoples here in Brazil, especially in the Amazon region, who still live, and these peoples that still live don't want to reach down to the level of survival.
>
> (UN Secretary-General 1987, 51)

This comment can be interpreted as a critique of contemporary perspectives on sustainability, including those related to global climate change, which run the risk of reducing the concept of "sustainability" to a matter of the biological survival of both humans and nonhumans.

Beyond the dictionary definition of the term, however, the word "life" when used as a descriptor of scholarly approaches carries a distinct nuance, at least in Japan. When Japanese social scientists opt for *seikatsu-ron*, or the life-centered approach, it is often because they feel that existing socioeconomic theories fall short of capturing the concrete realities of people's lives and fail to provide satisfactory answers to simple, yet urgent, questions such as "what we can actually do to live well?" (Torigoe 2002). This intellectual position can be traced back to the works of Norinaga Motoori in the Edo period, and scholars of folklore studies and rural sociology such as Kunio Yanagita (1875–1962) and Kizaemon Aruga (1897–1979). For Torigoe (2002), life-environmentalism is nothing but an application of this life-centered approach to environmental issues.

The interest of some Japanese scholars in the life-centered approach must be understood within the intellectual context of Japan, particularly since the beginning of the Meiji period. During this time, theories and concepts originating in western Europe were often regarded as superior and universal by many Japanese scholars, and their goal became how to properly comprehend them. In postwar Japan, for instance, there were significant debates among Japanese Marxist scholars regarding the "correct" interpretations of Japanese capitalism and the nature of "rural problems." Some of these debates can be seen as the extension of the pre-war debate between two Marxist schools of thought: the Kōza and Rōnō schools (Yasuba 1975). The social movements and policy making connected to these debates attracted much attention from the mass media, intellectuals, and a wide range of the public.

Yet, observers such as Tsuneichi Miyamoto (1907–1981), a scholar of folk cultures, believed that these debates and actions did not accurately reflect the realities of life in the countryside. In the early post-WWII period he observed:

> There was a rise of various active farmers' movements advocating for such things as land emancipation and land reforms. However, when I walked around agricultural regions ... I observed completely different dynamics at play. In fact, those [alternative movements] were more prevalent. They were deeply rooted in the realities on the ground, with people actively contemplating ways to increase production, build a healthy village, and recognize that they must join their hands together for success.
>
> (Miyamoto 1972, 249; author's translation)

Forceful social movements demanding land emancipation may seem like logical prescriptions from a Marxist perspective. However, Miyamoto's quote redirects our attention to those less vocal individuals who also aspired to improve their lives in ways that resonated with their own experiences. Life-environmentalism shares this perspective by recognizing that the thoughts, feelings, and aspirations of ordinary people are often overlooked or overly simplified. It emphasizes the importance of closely attending to their experiences in order to generate effective theories and practices.

Japanese scholars' affinity with the notion of *seikatsu* as a basic unit of analysis is also apparent in Torigoe's statement that "Japanese sociology traditionally concentrates mainly on people's lives or livelihoods rather than on society" (Torigoe 2014, 24). The idea of sociologists not concentrating on understanding "society" may sound oxymoronic, but consider the following observation by Callon and Law (1997):

> the very translation of Euro-American social thought into Japanese is extraordinarily difficult. For the whole idea of the "individual" and "society" is foreign to Japanese culture. There is a fascinating story to be retold about the conversion of these terms into Japanese neologisms—the ugly neologisms needed to import Euro-American social science and its problems into Japan. And another equally interesting story to be told of teaching about the distinction between the individual and society to eighteen-year olds in Japanese universities—students who tend to come from places which perform continuities between the collective and the personal, rather than divisions or dualisms. Are the Japanese disadvantaged? Perhaps. But perhaps not. For maybe what appears to be a Japanese problem is really one of Euro-American making.
>
> (166)

Callon and Law argue that the concepts of individual and society are inherently artificial. This realization led them to develop the actor–network theory, which acknowledges the interactions between both humans and nonhumans, and the emergence of agency from these interactions. The actor–network theory possesses certain qualities, such as its reluctance to attribute "essential characteristics" to any actor and its emphasis on exploring the specific and concrete aspects of phenomena (Ruming 2009), that resonate with life-environmentalism. While a comprehensive comparison is beyond the scope of this chapter, their recognition that "individual" and "society" are not self-evident units of analysis opens up possibilities for different assumptions upon which social scientific research can be based.

On the Notion of Community

While life-environmentalism finds the dualism of individual and society somewhat alien, the word "community" has found its way into its lexicon relatively easily. As Creed (2006) forcefully calls for a critical scrutiny of this "seemingly transparent term" (4), it is important to specify the origin, meaning, and nuance of the term in the life-environmentalism literature.

In Japanese, "community" is usually written phonetically as "コミュニティ" and sometimes used simply in place of the word, *kyōdōtai* ("common body"), or the German word, *Gemeinschaft*. While the Japanese word encompasses diverse meanings found in English, such as "communities" that are territorial or virtual, and vary in size from small to large, the "community" that life-environmentalism research typically focuses on refers to a territorially defined, relatively small group of people who engage in face-to-face interactions in everyday settings. This type of community aligns with the general characteristics of a "community" identified by Hillery (1955), including geographical boundaries, shared connections, and social interactions.

In premodern Japan, the concept of *"mura"* (agricultural settlement, hamlet, or village) exemplified this notion of "community" (Tsurumi 1975). In contemporary Japanese society, a public elementary school district often approximates the largest extent of these communities and mirrors the spatial boundaries of a *"mura"* in the premodern era. While some argue that small communities are diminishing in significance in the era of globalization, many life-environmentalist studies indicate that they are far from disappearing and, in some cases, are regaining their importance, particularly in the face of increasing natural disasters and socioeconomic challenges (e.g., Noda 2017; Kaneko 2017; Ueda and Torigoe 2012).

There are a few important considerations regarding the notion of "community" within the context of life-environmentalism. First, life-environmentalism research explicitly acknowledges that it does not assume communities to be homogeneous, coherent, or harmonious. Instead, the fundamental assumption is that any community can encompass internal diversity, fragmentation, conflicts, and contradictions. Therefore, community itself is not inherently categorized as "good" or "bad." However, we also recognize that the word "community" is sometimes used to serve as a signifier for specific actions and practices. For instance, a group of residents may refer to their activity of planting flowers along a street in their neighborhood as "community beautification." In such cases, "community" represents a concept that calls for the emergence of something yet to be defined, rather than something tangible (Gibson-Graham 2006, 413).

Second, another rationale behind the emphasis of life-environmentalism on the analysis of communities and community-level organizations, as opposed to families or households, is that communities are open to the broader society. According to Torigoe (2018),

> The reason life-environmentalism focuses on small communities is that, unlike families, which tend to be 'closed' to a private sphere, small communities are publicly open. To illustrate this, let's consider houses situated along a small river. Each house desires to freely utilize the water from the river. However, under the principle of "upstream houses (must act) for downstream houses," the community establishes regulations regarding wastewater management and water usage through the creation of specific rules. In other words, communities possess publicly open norms that are intended for the benefit of everyone.
> (534; author's translation)

The public nature of community rules and norms is important for life-environmentalism because it implies that these norms have some logical foundations that are, at least in principle, comprehensible for outsiders. Consequently, uncovering and comprehending these norms and rules is precisely what life-environmentalism research aims to achieve.

Finally, it is also important to note that within postwar Japanese regional development discourses and policies, the term "community" has been assigned a very specific meaning:

> By a "community" we refer to a group comprising individuals and families as independent and responsible citizens. This community is characterized by its local uniqueness, shared goals, openness, and mutual trust among its members.
> (Kokumin Seikatsu Shingikai 1969: authors' translation, 155–156)

This view of a community, articulated by a subcommittee of the Economic Planning Agency of Japan, essentially exhibits a modernist ideal, akin to the concept of "civil society," which, if realized, would replace the "old" communities that supposedly had feudalistic and even anti-modern characteristics (cf. Chen 2010). This modernist view of a community is clearly not the perspective of life-environmentalism. While life-environmentalists may not outright reject state-backed "community movements," they often emphasize that the "old" communities are not as feudalistic, hierarchical, and closed as they are often portrayed by media, politicians, and some scholars.[3]

How Does Life-Environmentalism Understand the World?

Placing "life" at the heart of socio-environmental analysis is far from new; it has been explored in various social scientific inquiries, both in the past and the present.[4] Therefore, the novelty of life-environmentalism must be sought in its distinct conceptualization and analysis of "life" and environment. In his 1989 book, Torigoe provides one of the first and most comprehensive theoretical overviews of life-environmentalism. Here, I offer my interpretation of Torigoe's discussion in order to provide a conceptual foundation to facilitate future dialogue with a contemporary English-language audience. I also draw insights from Torigoe's subsequent works and other life-environmentalism scholars. However, to better grasp the subtle nuances of Torigoe's arguments, readers are encouraged to refer to Chapter 18, and his original texts (e.g., Torigoe 1989, 1997, 2002).

In the following sections, I focus on six key ideas of life-environmentalism, namely "experience" (経験: *keiken*), emotional sensitivity (感受性: kanjusei), life consciousness (生活意識: *seikatsu ishiki*), the logic of persuasion (言い分: *iibun*), the handling of power issues, and life organization (生活組織: *seikatsu soshiki*). It is essential to note that these categories have emerged as practical necessities and conveniences in conducting intensive ethnographic fieldwork on environmental controversies, rather than as a result of theoretical or philosophical reflections (Torigoe 1989). Therefore, for social science researchers engaged in fieldwork, some of the assertions discussed here may seem like common sense. However, I believe that there are still distinct characteristics and emphases of life-environmentalism that are worth noting and exploring.

Experience (Rather than Action)

First, the focus on "experience" requires careful discussion as it is central to the thinking of life-environmentalism but is often misunderstood. Torigoe (1989) argues that when dealing with various environmental issues, the assumptions of rational actions, as defined by Weber's conceptions of rational–instrumental social action and rational social action with values, often hinder sound analysis of the issues. Instead, Torigoe (1989) says, "when it comes to the types of environmental problems we are concerned with, I do not find the action-based approach (*kōi-ron*) particularly effective. Therefore, I opt for the standpoint of the experience-based approach (*keiken-ron*)" (18; also Chapter 18, 261). We may call the latter the "experientialist approach."

But isn't an "experience" simply a "past action?" To answer this question, I often use in my classroom a fictional example of a block party, where residents of a neighborhood come together for a casual lunch. If one interprets this "action"—neighbors happily talking, for instance—by explicitly or implicitly assuming it to be a rational–instrumental social action, one may reason that the block party is a deliberate consequence of the residents' desire to socialize. Furthermore, if an observer asks one of the residents (Resident A) why he came to the party, he may respond, "because a party like this helps us to get to know each other better," reaffirming this instrumental interpretation (in this case, a block party facilitates socialization).

Emphasizing "experience" means not assigning excessive significance to the observed action or expressed opinion as a representation of underlying motives. Drawing on the imaginary example, life-environmentalists would consider the observed action of the resident as just one choice among many other possible actions under the particular circumstances. For example, the resident may have had various past encounters and events, such as:

- I've been friends with some of the neighbors.
- We have had disputes over Resident C's poorly maintained yard for many years.
- My parents always told me to go to these kinds of events.
- My spouse has been a close friend of Residents D and E.

Torigoe refers to the totality of these occurrences to Resident A that forms the basis for selecting his present and future actions from a range of possibilities as "experience." In this example, the resident may have decided to attend the party due to a small "trigger"—a call from one of his close friends the night before. However, Resident B had other options available, such as not attending at all, only staying for an hour, or sending his spouse in his place. If there had been a different "trigger," he might have chosen a different course of action. In such cases, his response to the question mentioned earlier might have also been quite different (e.g., "A block party is a waste of time").

Based on their field research experience, life-environmentalists have become acutely aware that local inhabitants often exhibit seemingly contradictory behaviors and opinions on different occasions. Torigoe (1989), for example, mentions that "we have encountered too many times the situation in which residents suddenly change their opinion" (21; also Chapter 18, 263). This realization forms the foundation of life-environmentalism's methodological emphasis on "experience" rather than "action." It also underscores the significance of studying the lived experiences and histories of the local community members, as exemplified in many chapters of this book.

As Torigoe readily admits, "The assertion that one must descend to the realm of experience in order to analyze the consequences and future possibilities of human actions is not particularly groundbreaking" (1989, 22; Chapter 18, 264). Nevertheless, at least in the 1980s, few studies put the methodological emphasis on "experience" on the forefront of socio-environmental analysis, and arguably even today many socio-environmental studies rely heavily on an instrumental interpretation of observed actions and on expressed opinions (although his does not mean that actions and opinions do not matter, as they do have real consequences).

Emotion and Mutual Non-Understanding

Another noteworthy characteristic of life-environmentalism is its explicit recognition of the significant role of affects and emotional sensitivity in shaping people's behaviors and actions. The emphasis on emotional sensitivity did not primarily stem from theoretical contemplation but rather from the demands of the "reality of the field" (Torigoe 1989, 24). The inclusion of emotion in social analysis is not unique, but what is intriguing in life-environmentalism research is the understanding of emotional sensitivity as a key factor contributing to various conflicts within a local community. Torigoe explains:

> I do not disregard the fact that differences in purpose (rational-purposeful) and values (value-rational) can also lead to serious conflicts. However, above all else, emotional sensitivity often causes the lack of mutual understanding in reality. In addition, it is surprising how often people's purposes and values originate from their emotional sensitivity. For example, it is not uncommon for individuals to start with the feeling that "nuclear power plants are frightening" and then construct a rational logic to align with the sentiment.
>
> (1989, 25; Chapter 18, 265)

This observation also leads to an important perspective, or core assumption, held by life-environmentalism. It suggests that the focus on local livelihoods does not imply the presumption of a cohesive and harmonious community. Instead, life-environmentalism assumes that a lack of mutual understanding (*sōgo murikai*; literally "mutual non-understanding") among local residents is the norm in the everyday lives of a community (Torigoe 1989, 1997).[5] A common insight derived from life-environmentalist analysis is that while environmental conflicts are often attributed to the absence of objective scientific knowledge or competing political ideologies, the root cause of these conflicts may actually lie in the realm of emotional sensitivity. This perspective also suggests the possibility that certain socio-environmental conflicts can be resolved or alleviated within local communities in everyday life, even in the absence of intellectual and ideological agreement among community members.

Scales of Everyday Life Consciousness

Third is the notion of *seikatsu ishiki*, literately translated as "life consciousness." However, the use of the word "consciousness" may be misleading as the English word has diverse meanings.[6] In the context of life-environmentalism, the term *seikatsu ishiki* refers to the views, feelings, values that a person or a group of people have about a given object and that inform their everyday actions. While experience forms the foundation of life consciousness, life consciousness also extracts and attributes meaning to specific experiences. Seen from this perspective, "life consciousness" is not an eccentric concept; rather, it is arguably a central concern in any ethnographic work.

Torigoe (1989) identifies three layers of life consciousness as analytically useful categories. The first layer is individually specific consciousness based on personal experiences (個人的な体験知: *kojinteki na taikenchi*). For example, someone who has experienced a major earthquake may develop an awareness that "earthquakes are scary." Such personal experiences and the resulting consciousness can significantly influence a person's future actions.

The second layer of life consciousness is shared among members of a community with a long history of living together. Torigoe (1989) refers to this layer of life consciousness as "living common sense" (生活常識: *seikatsu jōshiki*). This includes norms, perceptions, and values that are cultivated and sustained through various experiences, often outside of formal public education, in local communities with longstanding livelihoods. They are "not limited to techniques of production but also include the manners required towards leaders, seniors, the opposite sex and even ancestral spirits, as well as their role performance and ways to organize things" (Torigoe 2014, 30). While the "living common sense" of a particular community may appear anachronistic, parochial, or even irrational to outside observers such as government officials and scholars, life-environmentalism research often demonstrates that paying close attention to "living common sense" is crucial for understanding the nature of local conflicts and finding less conflict-ridden approaches to solving local socio-environmental problems.

The third layer of life consciousness is referred to as "popular morality" (通俗道徳: *tsūzoku dōtoku*).[7] Torigoe (2014, 29) describes it as "a life-consciousness inspired by the state in order to govern the people." He goes on to say, "Looking back at socialization during the era of modernization in Japan, a period of around 150 years, we must marvel at the power of national morality in which the public has been influenced by the Japanese government" (29). This popular morality instilled in the minds of many ordinary Japanese, most typically through public education, encompasses values such as diligence, saving, filial piety, and honesty. Similar notions centered around the nation-state are evident, with place-specific variations, in many Asian developmentalist states where the nation was the primary driver of modernization (Chen 2010).

Knowledge and the Logic of Persuasion

For Torigoe, life consciousness, the internal workings of people's minds, is articulated as a kind of knowledge that informs and rationalizes certain judgments and actions. Discursive logics of persuasion/justification, called *iibun* (literally, "a share of my/our say"), based on such knowledge play central roles at the site of socio-environmental problems and conflicts. For example, in the case of a dam construction project, several competing logics of persuasion are likely to emerge among the residents of the project site. One resident group may draw on a particular popular morality, claiming, "we must support the project in the name of the public good," while another group may base their argument on a different popular morality, stating, "this is a form of social and environmental injustice because…" Yet another group may appeal to the local knowledge and values (living common sense) and argue, "there is a sacred site in the project area, so we demand an alteration in the project."

In life-environmentalism research, the focus is typically on group-based discursive logics of persuasion, which may appear to pay insufficient attention to the diverse views and opinions of individuals. On this point Torigoe (1989) metaphorically sums up the reason as "We cannot understand the mind of a person, but we can understand the mind of a group of people" (45; also Chapter 18, 277 What does that mean?

Using the dam construction project as an example, when the construction of a dam is proposed, some residents may have a visceral fear of such a large-scale endeavor. Life-environmentalism researchers find it difficult to subject such bodily feelings, however real and significant they are, to analysis (hence "We cannot understand the mind of a person"). In practice, what typically happens is that residents form groups or alliances with like-minded individuals who share opinions and emotions. These groups then articulate a specific logic to persuade others as well as themselves, and justify their stance on the problem (i.e., dam project). "The mind of a group of people" is a metaphor of such a logic of persuasion.

Life-environmentalism chooses to study "the mind of a group of people" rather than the minds of individual persons for practical reasons. First, it recognizes the inherent difficulty in truly understanding an individual's inner world. Second, because the logic of persuasion (*iibun*) is typically grounded in some logical framework (e.g., not simply "we are scared," but "we are scared *because*..."), researchers can at least grasp the internal consistency of the logic. Third, and perhaps most importantly, life-environmentalism posits that individuals usually begin to align themselves with and act upon group-based logics of persuasion in response to various socio-environmental issues, rather than relying solely on discrete individual consciousnesses.[8]

The focus on *iibun* in life-environmentalism research helps us understand how it addresses the issues of affects and emotional sensitivity. Its distinctive characteristics become apparent when compared to certain strands of feminist scholarship that also acknowledge the significance of emotion. For example, Davis and Hayes-Conroy (2016) state:

> as researchers have noted in a variety of contaminated sites around the world, the emotional demands of bodies also serve as motivations for political organizing and actions that have changed, sometimes radically, the operation of state power and international markets in a variety of places.
> (Davis and Hayes-Conroy 2016, 126–127)

Here, the "emotional demands of bodies" of individuals are viewed as potential political forces to be scaled up to modify social structures. Essentially, emotions such as "fear" or "anger" are considered crucial sources of political action. While a life-environmentalist perspective does not necessarily deny the significance of emotion in those aspects, it posits that individual and collective emotions are initially expressed as a logic of persuasion to gain political traction. This apparent difference may serve as a potential opening for productive dialogue between life-environmentalism and other approaches focusing on emotion.

Power and the Two-Persons World

At first glance, life-environmentalism, with its focus on local everyday activities and their complexity, may appear to overlook the issues of power and politics, lacking explicit theorization of how class, gender, and capitalist relations shape local lives through exploitation, marginalization, and dispossession. Torigoe (1989), however, maintains that power is a central concern in life-environmentalism. According to Torigoe, the issue of power should be addressed as a problem of "morality," referring to the norms and values promoted by the state to control the nation.

Nevertheless, what sets life-environmentalism apart is its insistence on *not* analyzing environmental problems and conflicts from an objective, third person's perspective. While life-environmentalism may agree with the basic premise of political ecology that environmental change is a product of political processes (Robbins 2012), it distances itself from analyzing power relations from a third person perspective. Instead, it insists on analyzing socio-environmental issues from the standpoints of those being analyzed, typically those who live in a focal community, without resorting to the third person's perspective. Torigoe symbolically states that life-environmentalism analyzes issues in the world of "I and thou" (Torigoe 1989, 31).

As a result, power is not analyzed in an objective and abstract manner, focusing, for instance, on exploitative structures and power asymmetries. Instead, it is understood as concrete expressions observed through lived experiences. This epistemological stance of the "two-persons world" bears resemblance to an approach that emphasizes the value of focusing on the "social interface" (Long 2004; Turner 2012). Nevertheless, it is an open question whether empirical studies of life-environmentalism adequately address the issues of power from the perspective of those interested in articulating deeper causes of injustice and the nature of structural problems, such as certain strands of political ecological and environmental justice research.

Life Organizations and Environmental Change

In the analytical framework of life-environmentalism, "life organizations" or *seikatsu soshiki* often play a critical role. These organizations refer to local community-level groups and associations that deal with various aspects of everyday life activities. Examples of life organizations range widely, from relatively durable organizations like neighborhood associations (*jichikai*) and local festival associations, to groups formed specifically to address particular issues (e.g., "society of concerned residents over the dam project in Town X").[9] Life organizations are conceptually distinguished from productive/work organizations (*seisan soshiki*), which are organizations directly involved in the production of goods and services (e.g., agricultural cooperatives), although a single community organization may serve both roles in some cases.

Figure 2.2 provides a summary of the schematic process of how a local community responds to a demand for a change in their living environment, such as the proposed siting of an imposed locally unwanted facility, based on

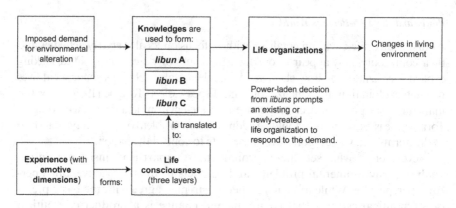

Figure 2.2 Conceptual flow leading to changes in living environment. Modified based on Torigoe (1989).

Torigoe (1989). In response to the proposed environmental alteration, different groups of inhabitants develop distinct logics of persuasion (*iibun*), drawing on knowledge based primarily on living common sense and popular morality to determine how to address it. According to the thinking of life-environmentalism, it is uncommon for groups with different *iibuns* to reach a complete consensus; instead, Torigoe emphasizes, the outcome is ultimately determined by the power relations among these groups.

In practice, the proposal for change is typically handled by one or more life organizations. In some cases, existing life organizations are utilized, while in other cases new life organizations may be established to cope with (e.g., resist, resolve, or adapt to) the proposal in the everyday lives of the community. As a result, these everyday lives themselves undergo changes over time, which in turn shape the experience of the inhabitants. This is why life-environmentalists pay close attention to the internal workings, employed logics, and ongoing transformations of these local life organizations, as demonstrated in many of the forthcoming chapters..

It is worth noting that life-environmentalism does not explicitly employ and emphasize the languages of sustainability or resilience. However, it can still be regarded as a framework that theorizes how a local community responds to various emerging pressures for change in living environment, whether it be related to natural disasters, locally unwanted facilities, or local revitalization projects, in order to sustain the well-being of their everyday life.

Notes

1 We use the hyphenated translation "life-environmentalism" rather than "life environmentalism," which has been a more common translation (e.g., Kada 2006; Torigoe 2014; Yamamoto 2019). This choice is made to emphasize the nuance that it is an ideological proposition (*shugi*) that prioritizes a particular kind of environment—the environment that is lived by people. Furthermore, other translations may carry the nuance of the Japanese term better, such as "living-environmentalism," "lived-environmentalism," or even "everyday-life environmentalism." However, because the

term "life environmentalism" has been already introduced and used in multiple publications, we decided to make only the minimal modification in this book.
2 The English word "life" has a broader range of meaning. In addition to referring to the "manner of living" as in *seikatsu*, it can also denote the state of being alive (*seimei*), the period from birth to death (*isshō*), biography (*jinsei*), and economic means of living (*seikei*). Each of these concepts has its own distinct words in Japanese.
3 For example, mass media and politicians frequently use *-mura* (village) to imply a closed circle of stakeholders (e.g., *genshiryoku-mura* or "nuclear village" composed of politicians, industry, and government-patronized scholars who promote the nuclear industry).
4 French geographer Paul Vidal de la Blache's *genres de vie* (lifeways) in the early 20th century and, more recently, the sustainable livelihood approach (Chambers and Conway 1991; Scoones 2015) are examples of approaches that center their analysis on life and livelihoods.
5 For the rationale to use the rather awkward translation, mutual non-understanding, see Chapter 18, footnote 9.
6 It is worth noting that *ishiki* is also a Buddhist term, a translation of Sanskrit, *mano-vijñāna* (mind discernment).
7 *Tsūzoku Dōtoku* (popular morality) is a term coined by the prominent Japanese historian Yoshio Yasumaru. It refers to a moral code that gained popularity among the common people, emphasizing diligence and frugality as the prime virtues. This moral code emerged during the late Edo period and continued into the Meiji period, a time when market-based economies were spreading and the wealth gap was widening. It is worth noting that *tsūzoku dōtoku* is not attributed to the modern state's creation, although it became a convenient means of governing the nation during the Meiji period. There is a clear connection to the notion of victim-blaming in contemporary society.
8 On one hand, the emphasis on group-based logic in life-environmentalism research reflects how local issues are typically addressed in Japan. Ueda and Torigoe (Chapter 9), for instance, demonstrate how the municipal office in the tsunami-hit area of Tōhoku was less inclined to respond to individual residents' requests for post-disaster support but showed greater willingness to assist when requests were made on behalf of resident organizations. On the other hand, we also take note that some English-language works in feminist political ecology emphasize the role of local collective environmental experience in shaping individual identities (Rocheleau, Thomas-Slayter, and Wangari 1996; Dey, Resurreccion, and Doneys 2013), cautioning against reducing one's identity to sociodemographic attributes such as gender, race, and class. This emphasis resonates with the focus of life-environmentalism on locally shaped, group-based consciousness.
9 Many life organizations, such as local funeral, festival, and religious groups, strongly embody the characteristics of *minjian* organizations (Chen 2010). In life-environmentalism research, the distinction between *minjian* and civil society organizations is not always clear. However, the life organizations that life-environmentalism research typically focuses on are local community-scale organizations that have direct connections to the everyday lives of local inhabitants. For example, a group of residents forming "Concerned citizens over the X dam project (in their community)" would likely be considered a life organization. On the other hand, an NGO named "Citizens against large hydro dams" may not be considered a life organization if it deals with the broader issues of hydro dams in general.

References

Callon, Michel, and John Law. 1997. "After the Individual in Society: Lessons on Collectivity from Science, Technology and Society." *Canadian Journal of Sociology/Cahiers Canadiens de Sociologie* 22 (2): 165–82. https://doi.org/10.2307/3341747

Chambers, Robert, and Gordon Conway. 1991. *Sustainable Rural Livelihoods: Practical Concepts for the 21st Century*. IDS Discussion Paper 296.
Chen, Kuan-Hsing. 2010. *Asia as Method. Toward Deimperialization*. Durham, NC: Duke University Press.
Creed, Gerald W., ed. 2006. *The Seductions of Community: Emancipations, Oppressions, Quandaries*. Santa Fe: School of American Research Press.
Davis, Sasha, and Jessica Hayes-Conroy. 2016. "Living with Contamination: Alternative Perspectives and Lessons from the Marshall Islands." In *Unravelling the Fukushima Disaster*, edited by Mitsuo Yamakawa, and Daisaku Yamamoto, 120–35. London: Routledge.
Dey, Soma, Bernadette P. Resurreccion, and Philippe Doneys. 2013. "Gender and Environmental Struggles: Voices from Adivasi Garocommunity in Bangladesh." *Gender, Place & Culture* 21 (8): 945–62. https://doi.org/10.1080/0966369X.2013.832662
Furukawa, Akira. 2007. *Village Life in Modern Japan: An Environmental Perspective*. Melbourne: Trans Pacific Press.
Gibson-Graham, J.K. 2006. *A Postcapitalist Politics*. Minneapolis, MN: University of Minnesota Press.
Hillery, G.A. 1955. "Definitions of Community: Areas of Agreement." *Rural Sociology* 20 (2): 111–23.
Kada, Yukiko. 2006. "Three Paradigms behind River Governance in Japan: Modern Technicism, Nature Conservationism and Life Environmentalism." *International Journal of Japanese Sociology* 15 (1): 40–54.
Kaneko, Hiroyuki. 2017. "Radioactive Contamination of Forest Commons." In *Unravelling the Fukushima Disaster*, edited by Mitsuo Yamakawa, and Daisaku Yamamoto, 136–53. London: Routledge.
Kokumin Seikatsu Shingikai. 1969. *Komyunitei: Seikatsu no Baniokeru Ningensei no Kaifuku* [Community: The Restoration of Humanity in the Place of Living]. Tokyo: Economic Planning Agency of Japan.
Miyamoto, Tsuneichi. 1972. *Mura No Hōkai* [Destruction of Villages]. Tokyo: Miraisha.
Noda, Takehito. 2017. "Why Do Local Residents Continue to Use Potentially Contaminated Stream Water after the Nuclear Accident?" In *Rebuilding Fukushima*, edited by Mitsuo Yamakawa, and Daisaku Yamamoto, 53–68. London: Routledge.
Robbins, Paul. 2012. *Political Ecology: A Critical Introduction*. Second edition. Chichester, West Sussex: John Wiley & Sons.
Rocheleau, D., B. Thomas-Slayter, and E. Wangari. 1996. *Feminist Political Ecology: Global Issues and Local Experience*. Global Issues and Local Experience. London: Routledge.
Ruming, Kristian. 2009. "Following the Actors: Mobilising an Actor-Network Theory Methodology in Geography." *Australian Geographer* 40 (4): 451–69. https://doi.org/10.1080/00049180903312653
Scoones, Ian. 2015. *Sustainable Livelihoods and Rural Development*. Rugby, UK: Practical Action Publishing. https://doi.org/10.3362/9781780448749
Torigoe, Hiroyuki, ed. 1989. *Kankyō Mondai No Shakai Riron: Seikatsu Kankyō Shugi No Tachiba Kara* [Social Theory of Environmental Problems: From the Standpoint of Life-environmentalism]. Tokyo: Ochanomizu Shobō.
———. 1997. *Kankyō Shakaigaku No Riron to Jissen: Seikatsu Kankyō Shugi No Tachiba Kara* [Theory and Practice of Environmental Sociology: Perspectives of Life Environmentalism]. Tokyo: Yūhikaku.

———. 2002. *Yanagita Minzokugaku no Firosofi* [English title: *Yanagida Kunio and Japanese Folklore*]. Tokyo: University of Tokyo Press.
———. 2014. "Life Environmentalism: A Model Developed under Environmental Degradation." *International Journal of Japanese Sociology*, 23 (1): 21–31.
———. 2018. *Genpatsu Saigai to Jimoto Komyuniti: Fukushima Ken Kawauchi Mura Funtōki* [Nuclear Disaster and a Local Community: Records of Struggles in Kawauchi Village, Fukushima]. Tokyo: Tōshindō.
Torigoe, Hiroyuki, and Yukiko Kada, eds. 1984. *Mizu to Hito no Kankyōshi: Biwako Hōkoku Sho* [Environmental History of Water and People: Lake Biwa Report]. Tokyo: Ochanomizu Shobō.
Tsurumi, Kazuko. 1975. "Yanagida Kunio's Work as a Model of Endogenous Development." Tokyo: Institute of International Relations, Sophia University.
Ueda, Kyoko, and Hiroyuki Torigoe. 2012. "Why Do Victims of the Tsunami Return to the Coast?" *International Journal of Japanese Sociology*, 21 (1): 21–9.
UN Secretary-General. 1987. "Report of the World Commission on Environment and Development: Our Common Future." New York: United Nations. https://digitallibrary.un.org/record/139811
Yamamoto, Daisaku. 2019. "Life Environmentalism." In *Wiley Blackwell Encyclopedia of Sociology*, edited by George Ritzer. Oxford, UK: John Wiley & Sons.
Yasuba, Yasukichi. 1975. "Anatomy of the Debate on Japanese Capitalism." *Journal of Japanese Studies* 2 (1): 63–82. https://doi.org/10.2307/132039

Part I
Developmental Impulse and Everyday-Life Organizations

3 Local Rules

Sustaining Local Everyday Life with Aqua-Tourism

Takehito Noda

The global pandemic of COVID-19 has raised critical question about the conventional approaches to tourism (Lew et al. 2020). Tourism is a vital component of contemporary globalization, and has accelerated the movement of people, goods, and money across the planet. Indeed, global tourism played a significant role in the rapid and widespread transmission of the virus, first identified in Wuhan, China. Nevertheless, even before the outbreak, issues of overtourism were already prevalent in various destinations worldwide. What should the future of tourism look like in the post-COVID era? This chapter examines a local practice of aqua-tourism in Japan as a means to envision a more sustainable form of tourism in the post-pandemic world.

Since the 1990s, research on "sustainable tourism" has grown considerably. While some proponents of sustainable tourism focus primarily on the sustainability of the tourism industry itself, most would agree that tourism should serve as a pathway toward achieving or transitioning to a more sustainable form of development (e.g., Hunter 1995; Aall 2014). However, the question of "sustainability of what and for whom" remains a topic of ongoing debate. Some emphasize the sustainability of the ecological environment (Dubois and Ceron 2006), while others prioritize resource management (Carter, Baxter, and Hockings 2001) or social and cultural aspects (Hardy, Beeton, and Pearson 2002).

While each of these focuses has its own merits, this chapter takes a stance that emphasizes the sustainability of local community life. It is worth noting that a community-focused approach is not new or unique in tourism research (e.g., Murphy 1985; Blackstock 2005; Ishihara 2020). However, as I will clarify, many studies on community-based tourism primarily focus on community participation in tourism governance and management, rather than the intersection of tourism and the everyday lives of local communities.

In this chapter, I adopt a life-environmentalism perspective to analyze local aqua-tourism in Japan, with a particular emphasis on the sustainability of residents' everyday life. Here, "aqua-tourism" broadly refers to various tourism activities that take place near bodies of water, such as lakes, rivers, oceans, and springs (Ishimori 2011). Therefore, activities like visiting waterfalls, swimming in the sea, kayaking, scuba diving, and sport fishing all fall under the umbrella

DOI: 10.4324/9781003185031-4

of aqua-tourism. What distinguishes Japanese aqua-tourism is the existence of many cases where private or communal spaces that were originally part of the everyday life of local residents (e.g., springs and washing areas) have gradually transformed into tourism resources. In other words, the everyday spaces of local residents have become objects of tourist gaze (Urry 1990; MacCannel 1999). Due to this nature of aqua-tourism, which takes place in the everyday life spaces of residents, it is essential to explore ways of supporting tourism that residents can genuinely embrace.

Contemporary Challenges of Aqua-tourism

Figure 3.1 shows a communal *mizuba* ("water area") in the Ikuji district of Kurobe City, Toyama Prefecture. This natural spring is used by local residents for obtaining water, washing vegetables, and laundering clothing. Users who frequent this water area likely have access to washing machines at home, but they still choose to come here, stating, "because you can meet someone here." This water area is open to tourists, attracting an average of 105 visitors per day (Onozawa, Yoshizumi, Suzuki et al. 2008). However, the presence of tourists in such water spaces can potentially lead to conflicts between residents and tourists. Tourists are not attracted to the clean natural water flowing out from the springs; instead, they are fascinated by the ways water is integrated into the

Figure 3.1 Water area where residents and tourists co-use. Photograph by the author.

residents' everyday lives and the accompanying scenes. These scenes include water being drawn into private kitchen spaces and residents washing clothes together with fellow residents.

Therefore, the coexistence of tourism and residents' lives is an urgent issue for aqua-tourism in these circumstances, and there are currently two major policy directions in response. The first is simply to demand better behavior from tourists. For example, in Kyoto, a world-famous tourist city, the issue of "maiko paparazzi" has been a major concern in the Gion district, where tourists chase and take photographs of *geiko* (also known as *geisha*) and *maiko* (apprentice *geiko*) on their way to work. In response, pictogram signs were created to communicate the rules and manners to be observed in the area (Figure 3.2). More typically, various tourist destinations suffer from "tourism pollution," including noise, littering, and congestion, and conflicts between hosts and guests (Smith 1977). In the case of aqua-tourism, where residents' everyday living

Figure 3.2 Signboard for tourists in the Gion District of Kyoto. Photograph by the author.

spaces are subjected to the tourist gaze, these problems are often felt more intensely by the local communities.

The second policy direction is the spatial separation of tourist resources and residents' everyday lives. In the context of the aqua-tourism discussed here, a close proximity of residents and tourists creates a high likelihood of conflict between them. One possible response is to create dedicated water areas where tourists can freely visit and enjoy. This approach is frequently seen in Japan (Hashimoto 1999). However, in reality, tourists rarely visit and use such designated water areas. For instance, in Matsumoto City, Nagano Prefecture, which is known for aqua-tourism, there is a water area with natural spring water built for tourists in the public square in front of the main train station. Yet, it is rare to see any tourists there (Figure 3.3). The reason for this is quite evident. Tourists are attracted to water areas that residents actually use in their daily lives and that have a sense of everyday use. Not only does the water area shown in Figure 3.3 look like an unattractive monument, but such designated water areas also tend to lack maintenance and risk deteriorating water quality due to the absence of continuous use and monitoring by users.

In these ways, the limitations of the currently popular policy approaches are becoming increasingly evident. However, if no measures are taken and the current situation is left unaddressed, the issue of overtourism is likely to worsen

Figure 3.3 Water area for tourists in front of the JR Matsumoto station. Photograph by the author.

and may result in various problems for these local communities. Indeed, if the touristification of these water areas remains uncontrolled, it could lead to the complete abandonment of their local use. Consequently, the appeal as a tourist destination would also diminish. The question, therefore, is how to make these water areas accessible to tourists without depriving local residents of their regular use.

Local Rules Supporting the Commons

To answer the question, we must first examine why aqua-tourism can lead to discords and conflicts within local communities. Conventional views attribute the reason to bad manners or to inadequate tourism management (e.g., insufficient separation of tourist and resident spaces). I propose instead that the underlying reason lies in the fact that the object of tourism is a local water commons (Ostrom 1990). For instance, the water area in Figure 3.1 may be viewed simply as a natural resource where water springs out. However, it is actually a local commons maintained by local use and management rules established by the Washing Area Management Association, an organization formed by the residents of the community (Noda 2018).

Examining the water area closely, we can see that it consists of the main water source in the back, where water spring out from the ground, and five connected stainless steel water tanks. The top tank is for drinking water, the second is for cooling vegetables or bottled drinks, and the lower tanks are for washing purposes. In essence, the upstream water is reserved for "cleaner" uses and the downstream water for "dirtier" ones. When cutting fish, the user can turn the faucet 180 degrees and drain the used water into the waterway on the other side.

There are also rules for the maintenance of the water area. Users clean the area, including the spring, every Saturday from 8:00 a.m. for approximately 30 minutes. Participation in the cleaning activity is mandatory, and a rotation system is implemented where a group of three people, representing different three households, takes turns each week. Specific procedures for cleaning are followed, and special care is taken to prevent rapid algae growth inside the tanks. By the end of each week, if algae are visibly growing, those who cleaned last may be teased for not doing a thorough job. Residents consider cleaning skills as reflective of one's character and not something to be taken lightly.

In this case, only those who fulfill the labor obligation of cleaning are granted the right to use the water area. The right to use and the obligation to maintain are inseparable. If someone consistently neglects their cleaning duty, they will be deprived of the right to use the area. It is precisely because of these local rules and norms regarding use and maintenance that the quality of the water area has been maintained over many years.

What happens when this kind of usage and management system breaks down due to touristification? In a town in the Tōhoku region known for its spring water, local residents used to regularly use the water area and perform periodic cleaning. However, when the local government began to promote aqua-tourism in the region, it decided to outsource the cleaning of the water area to a specialized company in an effort to alleviate the burden on residents. To the local government's surprise, the water area became practically abandoned and was no longer used by the residents. Weeds started to grow, and pools became infested with algae. Naturally, tourists were not interested in a silent water area that lacked resident activity. Consequently, the number of tourists quickly declined, and local residents also expressed dissatisfaction with the water area's diminished accessibility. This was because the residents had established their right to use the water area by taking responsibility for regular cleaning. Despite the well-intentioned motives of the local government, when the cleaning company assumed the cleaning duties, the residents felt that their right to use the area had also been taken away.

It is important to note that this sense of entitlement and the rules governing the commons are not always explicitly recognized by the users. Rather, they often become apparent when they are infringed upon or when conflicts arise regarding the commons. This suggests in turn that if these local rules are carefully observed and adapted for tourism purposes, genuinely appealing tourist destinations may be created and sustained.

How Can the Degradation of Tourist Areas Be Prevented?

Let us consider the case of the Harie District in Takashima City, Shiga Prefecture as an example of "successful" aqua-tourism, where the residents have been able to preserve the unique character of their community and prevent the degradation of their water-based commons. Harie Village is a small lakeside community situated in the northwestern part of Lake Biwa. Prior to the pandemic, it attracted nearly 10,000 visitors annually. Interestingly, the community seems hesitant to embrace its status as a tourist destination, as evidenced by a sign on th community message board stating, "this is not a tourist area" (Noda 2013, 2014). This apparent contradiction, where the community welcomes visitors but does not identify itself as a tourist area, makes Harie an intriguing example of local aqua-tourism practice.

Aqua-Tourism as a Community Business

The Harie District in Takashima City, Shiga Prefecture, is on an alluvial fan created by the Azumi River that flows through the area into Lake Biwa. Due to its location, natural spring water emerges when wells are dug. Local residents have referred to this water as *shozu* (生水), meaning "living-water." They draw the water into their private properties and use it in a space called *kabata*, which practically serves as an extended kitchen area, often found in separate small huts. The water is used for drinking, washing vegetables, cooling fruits, raising carp, and cleaning dishes. Out of the approximately 170 households (comprising 660 people) in the district, 110 households make use of *kabata*.

A *kabata* consists of two or three interconnected pools, each serving a specific purpose (Figure 3.4). The first pool, called *motoike* ("source pond"), has gushing water and is used for drinking, cooking, and washing. The second pool, known as *tsuboike* ("pot pond"), is used for rinsing and refrigerating vegetables. The third pool, called *hataike* ("end pond"), functions as a washing area for dirty items. Carp are released into the hataike, and when pots or kettles used for cooking are soaked, the carp consume leftover grains of rice and food particles, helping to keep the water relatively clean. This *kabata* system is what attracts a significant number of tourists.

In Harie, there are two organizations responsible for managing water resources, including *kabata*, in the community. The first is the local *jichikai* (neighborhood association) called Harie-Ku, which coordinates an annual waterway cleaning activity involving all households, as well as quarterly cleaning of the Harie Ōkawa River. The second organization is the *Harie Shōzu no Sato Iinkai* (Committee on Harie Water Village, hereafter the "Committee"), established in 2004. This community-level nonprofit organization (NPO) handles various environmental conservation activities and community issues that the *jichikai* alone cannot address.

The Committee was initially formed to address emerging issues after the *kabata* was featured on a Japanese public television program, leading to an increase in visitors to the community. As *kabata* serves as the residents' private kitchen space, concerns arose with the growing number of visitors occasionally entering these areas. In response, the Committee developed a guided tour program to manage such issues. Local volunteer guides lead visitors through approximately ten *kabatas* during the tour, charging a fee of 1,000 yen (about US$9) per visitor. This generates annual revenue of nearly 10 million yen, which is used for various community activities and projects. In essence, aqua-tourism in Harie is a community-driven enterprise that emerged to address local everyday problems rather than being primarily profit-oriented.

Nevertheless, initially, gaining approval for the Committee's activities from the residents was difficult. From the residents' perspective, the Committee's efforts appeared to be focused on pro-tourism activities aimed at profit-making. While there were no open conflicts, a discord started to emerge between the Committee and some residents, including the *jichikai*, regarding the nature of aqua-tourism organized by the Committee. This discord arose because the Committee's activities seemed to infringe upon the local rules governing water resources in the community.

Local Rules on Water Resources

When the Committee members realized that they did not have full support of the community, they began to listen more attentively to the voices of the residents. It became evident that some residents perceived issues with the Committee's apparent profit-seeking activities that used the "nature" long maintained by the local community.

Figure 3.4 Kabata area in the Harie District. Photograph by the author.

To address these concerns, the Committee made explicit efforts to reinvest the economic benefits derived from their activities back into the local community. For instance, they donated photocopy machines and air conditioners to the local public hall and senior citizens' association, and covered the costs of food at local festivals and sports days. These expenses were typically covered by the *jichikai*, which collects fees from residents. The Committee prioritized equality in distributing benefits across the community, striving to ensure fairness. Sharing economic benefits in this manner should have been regarded as a respected response aligning with the principles of equality and fairness. However, instead of gaining understanding or applause from residents, the Committee's actions were met with criticism within the community. The reason for this must be understood by examining the rules governing water resources, including *kabata*, in the community.

The basic rules are as follows: each household is allowed to install one *kabata* for drinking water and daily use, but setting up a *kabata* for commercial purposes, such as selling water in plastic bottles, is prohibited. The *kabatas* in the neighborhood are interconnected through waterways, and drainage from upstream *kabatas* flows to downstream *kabatas*. Consequently, it is natural and essential for upstream households to consider their water usage in relation to downstream households.

Regarding maintenance, as previously mentioned, all households are obligated to participate in the annual waterway cleaning and quarterly cleaning of the Harie Ōkawa River as part of *jichikai* events. All drainage water from each household's *kabata* eventually flows into the river. Therefore, residents understand that fulfilling these cleaning obligations helps keep their *kabatas* in good working order. In other words, even though water springs up on private property, it is not something that can be used arbitrarily by the owner; it belongs to the commons of the community. In these ways the right to use and the obligation to maintain the water resource are considered an inseparable set under the local rules.

Viewed from this perspective, it becomes clear why the monetary contributions and donations made by the Committee became the target of criticism. The guided tours organized by the Committee led to new and increased utilization of the water resource by visitors. However, according to the community's rules, the use of *kabata* is permitted only if the maintenance obligations are upheld and cannot be simply bought with money. Although the Committee members were likely aware of the local rules, they may not have fully realized how important these rules were in the minds, or the life consciousness, of the residents. In fact, the residents themselves may not have recognized the significance of the rules until this event occurred. Hence, the criticisms came as a great surprise to the Committee.

Nonetheless, the Committee took the criticisms seriously and responded by actively participating in and leading maintenance activities related to water resources. For example, a small inner lake at the lower reaches of the Harie Ōkawa River had not been properly cleaned for a long time due to a lack of workers, and aquatic plants had overgrown, covering the water's surface. The Committee decided to take responsibility for cleaning the inner lake. However, as the Committee alone did not have sufficient manpower, they sought volunteers from outside the community. To attract volunteers, they proposed a work-tour that combined kabata cleaning and volunteer work. Although participants were required to pay for this work-tour, approximately 100 participants were secured in no time.

To the Committee's surprise, residents also criticized this initiative. One resident said, "I cannot believe that there are people who want to pay to clean the river, which we are only grudgingly doing. If that's the case, let those outsiders do all the work." Such voices alerted the Committee members. They realized that the cleaning of the Harie Ōkawa River was meant to be a collective effort involving all community members. Regardless of good intentions, the Committee's involvement of nonlocal participants in the cleaning infringed upon the rules governing the management of water resources in the community.

From that point forward, the Committee began seeking alternative ways to enhance the well-being of the community. They worked more closely with the *jichikai*, with whom they previously had a less favorable relationship, while also adhering to the local rules. These efforts included releasing carp into the waterways, installing planters along the waterways, and managing neglected bamboo forests in the area. Although these activities might appear insignificant or cosmetic, they gradually gained high regard from residents due to their contribution to a community that was aligned with the (previously unconscious) local rules governing water commons. Through these endeavors, the *kabata* tour itself came to be firmly understood as an activity aimed at protecting the everyday life of the community rather than solely for profit-making. This is how aqua-tourism gradually gained acceptance within the local community.

Local Rules as Barriers to the Vulgarization of Tourist Destinations

The local rules of the Harie District also serve as a safeguard against the vulgarization of the tourist area. Since all the *kabatas* in the community are located on private property and used by the residents in their daily lives, tourists often request access to water without intruding on private properties. In response, during a Committee meeting, a male participant proposed the construction of a new *kabata* where visitors could freely drink or draw spring water. Initially, this idea seemed sensible as it would alleviate the burden on residents who were involved in showcasing their *kabatas*. However, all the female members of the Committee strongly opposed the proposal, leading to its abandonment.

The women who opposed the idea explained their reasoning, stating that they were taught by their parents not to take *kabatas* lightly and that *kabatas* were considered sacred sites where gods resided. They also likened the situation to their approach to fishing, stating that "just as we can take the lives of fish only for eating, not for fun, we should be allowed to build a *kabata* only for the needs of our everyday life." Eventually, the male members of the Committee were convinced by these views and acknowledged that they had nearly forgotten the importance of such considerations. Consequently, the decision was made to forgo the installation of a tourist-oriented *kabata*.

As mentioned earlier, in many regions promoting aqua-tourism, new water-based facilities are often constructed to enhance convenience for tourists and alleviate the burden on residents. However, tourists are seldom attracted to such facilities as they perceive them merely as monuments. The construction of tourist-oriented facilities may seem like an ideal solution for balancing the desires of tourists and the daily lives of residents, but in reality, it can diminish the appeal of tourist destinations.

In contrast, the Harie District, despite all the twists and turns, has succeeded in preventing the vulgarization of its tourist area by upholding, and in some cases discovering, the local rules regarding water resources and applying them to tourism. In Harie, despite occasional requests by tourists, no souvenir shops have been built, and water is not sold. These decisions are made by referring to the local rules regarding the water resources in the community. From this perspective, these local rules, which have operated visibly or somewhat covertly within the community's history, are not obstacles to tourism development. Instead, they serve as a powerful means to enhance the community's unique character.

Local Rules as a Means of Sustainable Tourism

Drawing on the cases of aqua-tourism in Japan, this chapter has focused on the issues of sustainability of local everyday lives and tourism development. The water resources that this chapter examines are the local commons that have been supported by various local rules, some of which are not explicitly articulated or formalized *a priori*. For local residents, the right to use and the obligation to manage water are inseparable. Accordingly, policies to develop aqua-tourism must find a way to "open up" water sources to tourists without violating these local rules that define proper ways to use and maintain the natural resources.

Conventional policy measures employed in aqua-tourism have often aimed either to enforce better behavior from tourists or to spatially separate tourists from the residents' way of life. Instead, I shed light on an alternative approach where closely observing local rules and applying them to tourism can protect and enhance the sustainability of residents' everyday life, thereby also increasing the sustainability of tourism

Observing the local rules of a community is crucial for two reasons. First, they help shape the unique character of the community while preventing its vulgarization as a tourist area. Second, these local rules reflect the values and experiences of the local community, making them a means for tourists to learn about the value and significance of water resources for the local community. The water resources involved in aqua-tourism are not simply natural objects but may be seen as sacred entities, local commons, or social resources that reflect the everyday life consciousness of the residents. This is precisely what makes aqua-tourism appealing.

The 21st century is said to be the era of tourism. In order to realize sustainable tourism, the United Nations and the World Tourism Organization (UNWTO) have adopted the Global Code of Ethics for Tourism to improve the behavior and etiquette of tourists. Discussions around ethical tourism are emerging, emphasizing the demand for ethical practices (MacCannell 2011; Lovelock and Lovelock 2013). However, as demonstrated in this chapter, simply focusing on "tourism ethics" and demanding better manners from tourists is not a fundamentally viable solution for sustainable tourism when residents' livelihood resources are impacted. Given our interest in diverse forms of tourism practices around the world, it is crucial to promote tourism that respects and embraces the local rules operating in each area.

In the field of sustainable tourism literature, there is a growing number of studies that focus on local host communities and the livelihoods of residents (e.g., Wu and Pearce 2014; Su et al. 2016; Su et al. 2019). However, not many of these studies explore deeply into the values and norms of local communities, including the investigation of local rules that are not always codified or readily recognized even by the residents themselves. This is where the perspectives of life-environmentalism can offer unique and valuable insights, and this chapter has provided as an empirical example of this approach in the context of contemporary tourism.

References

Aall, Carlo. 2014. "Sustainable Tourism in Practice: Promoting or Perverting the Quest for a Sustainable Development?" *Sustainability (Basel, Switzerland)* 6 (5): 2562–83.
Blackstock, Kirsty. 2005. "A Critical Look at Community Based Tourism." *Community Development Journal* 40 (1): 39–49.
Carter, R. W., G. S. Baxter, and M. Hockings. 2001. "Resource Management in Tourism Research: A New Direction?" *Journal of Sustainable Tourism* 9 (4): 265–80.
Dubois, Ghislain, and Jean-Paul Ceron. 2006. "Tourism and Climate Change: Proposals for a Research Agenda." *Journal of Sustainable Tourism* 14 (4): 399–415.
Hardy, Anne, Robert J. S. Beeton, and Leonie Pearson. 2002. "Sustainable Tourism: An Overview of the Concept and its Position in Relation to Conceptualisations of Tourism." *Journal of Sustainable Tourism* 10 (6): 475–96.
Hashimoto, Kazuya. 1999. *Kankō Jinruigaku no Senryaku – Bunka no Urikata, Urarekata* [Strategies in Tourism Anthropology: How to Sell Culture, and How it is Sold]. Kyoto: Sekaishisō-sha.
Hunter, Colin J. 1995. "On the Need to Re-Conceptualise Sustainable Tourism Development." *Journal of Sustainable Tourism* 3 (3): 155–65.
Ishihara, Yusuke. 2020. "Overview of Community-Based Tourism." In *The Routledge Handbook of Community-Based Tourism Management*, 1st ed., 26–38, edited by Sandeep Kumar Walia. United Kingdom: Routledge.
Ishimori, Shūzō. 2011. "'Mizu no Wakusei' ni Okeru Kankō" [Sightseeing on the Water Planet] *Mahora* 68: 8–13.
Lew, Alan A., Joseph M. Cheer, Michael Haywood, Patrick Brouder, and Noel B. Salazar. 2020. "Visions of Travel and Tourism after the Global COVID-19 Transformation of 2020." *Tourism Geographies* 22 (3): 455–66.
Lovelock, Brent A., and Kristen M. Lovelock. 2013. *The Ethics of Tourism: Critical and Applied Perspective*, London: Routledge.
MacCannell, Dean. 1999. *The Tourist: A New Theory of the Leisure Class*, 3rd ed. Oakland, CA: University of California Press.
———. 2011. *The Ethics of Sightseeing*, Oakland, CA: University of California Press.
Murphy, Peter E. 1985. *Tourism: A Community Approach*. Oxford: Routledge.
Noda, Takehito. 2013. "Community Conflict Brought by Community Tourism." *Journal of Rural Studies* 20 (1): 11–22.
———. 2014. "The Meaning of Non Economic Activities in a Community Business: Water Resources Utilized by Tourism in Harie Village, Takashima City, Shiga." *Kankyō Shakaigaku Kenkyū* [Journal of Environmental Sociology] 20: 117–32. In Japanese.
———. 2018. "Komonzu no Haijosei to Kaihōsei– Akitaken Rokugō Chiku to Toyamaken Kiji Chiku no Akuatsūrizumu e no Taiō kara" [Exclusivity and Openness of Commons: From Responses to Aqua Tourism in Rokugo District, Akita Prefecture and Ikuji District, Toyama Prefecture]. In *Seikatsu Kankyō Shugi no Komyuniti Bunseki* [Community Analysis of Life-Environmentalism], edited by Hiroyuki Torigoe, Shigekazu Adachi, and Kiyoshi Kanebishi, 25–43. Kyoto: Minerva Publishing.
Onozawa, Michiko, Yoshizumi Yūko, and Suzuki Takeshi. 2008. "Araiba no Jizokuteki Riyō to Sono Henyō ni Tsuite no Kenkyū–Kurobeshi Senjōchi Yūsuigun Kiji Chiku no Shimizu o Jirei to shite" [A Study on Sustainable Use of Washing Areas and Their Transformation: A Case Study of Spring Water in the Kurobe River Alluvial Fan

Spring Area]. *Nihon Kenchiku Gakkai Kinki Shibu Kenkyū Hōkokushū. Kenchikukei* [Architectural Institute of Japan Kinki Branch Research Report] 48: 333–6.

Ostrom, Elinor. 1990. *Governing the Commons: The Evolution of Institutions for Collective Action.* Cambridge, UK: Cambridge University Press.

Smith, Valene, ed. 1977. *Hosts and Guests: The Anthropology of Tourism.* Philadelphia: University of Pennsylvania Press.

Su, Ming Ming, Geoffrey Wall, and Kejian Xu. 2016. "Heritage Tourism and Livelihood Sustainability of a Resettled Rural Community: Mount Sanqingshan World Heritage Site, China." *Journal of Sustainable Tourism* 24 (5): 735–57.

Su, Ming Ming, Geoffrey Wall, and Yanan Wang. 2019. "Integrating Tea and Tourism: A Sustainable Livelihoods Approach." *Journal of Sustainable Tourism* 27 (10): 1591–608.

Urry, John. 1990. *The Tourist Gaze: Leisure and Travel in Contemporary Societies.* London: Sage Publications.

Wu, Mao-Ying, and Philip L. Pearce. 2014. "Host Tourism Aspirations as a Point of Departure for the Sustainable Livelihoods Approach." *Journal of Sustainable Tourism* 22 (3): 440–60.

4 Coexistence without Consensus

The Role of a Life Organization in Mediating Between Fishermen and Surfers in a Coastal Community

Shusuke Murata

Until its discovery as a surfing spot by U.S. soldiers stationed in the area in the 1960s, the town of Kamogawa had been a quiet fishing port along the coast of Chiba Prefecture. Since that time, fishing and surfing in the community have had rather different trajectories. On the surfing side, Kamogawa evolved into a mecca for Japanese surfers, drawing migrants and weekend warriors from the nearby Tokyo metropolis. On the fishing side, the outflow of local youth population during the country's rapid economic growth through the late 1960s to the 1980s continues to challenge the industry. Faced with these challenges, in the late 1990s the fishing sector in Kamogawa began hiring surfers who had relocated to the area for warm water and waves on their boats as fishermen. Today, these in-migrant surfers constitute a large portion of the fishing population of Kamogawa. Since most fishing communities in Japan are facing a serious shortage of labor, this case of in-migrant surfers becoming fishermen has become a well-known example of how local labor shortages can be creatively resolved.

Yet the oft-reported stories of harmonious resolution do not accurately capture the reality of the situation in Kamogawa. The fact is that fishermen and surfers have often come into conflict over the utilization of the marine environment. These conflicts boil down to the question of whether Kamogawa is a fishing or surfing town. While fishermen despise large waves as a liability to safe and efficient fishing operations, surfers relish them. The resulting conflict between these completely contrasting views of waves culminated in a hotly contested mayoral election in the early 1990s in which fishermen and surfers supported their respective candidates. To overcome this conflict and reach consensus the local government held a series of meetings to discuss the use of coastal and marine resources. A volunteer citizen group also held an international conference, inviting nature conservationists and prominent surfers from around the world to discuss these issues. However, these venues aimed at "democratic consensus-building" gradually waned without any meaningful outcomes, and the disagreements between fishermen and surfers in Kamogawa persist today.

Notions of "democratic consensus-building" assume that it is essential for opposed parties to clearly elucidate their own position and come to terms with the other side's position in order to resolve conflicts that arise within a community. In Kamogawa, however, fishermen and surfers remain embroiled in conflict over their contrasting visions of the marine environment while also managing, at the same time, to somehow live and work together in the community. This fact raises the question of how these opposed parties manage to reconcile conflict and consensus. This chapter aims to answer this question by focusing on the details of the experience of fishermen and surfers in order to illustrate the mechanisms that enable their coexistence at the level of everyday life. To preview the findings of the chapter, the local community and the fishing industry embraced the surfers who moved to Kamogawa in search of waves as community members; some of the migrant surfers are now "surfing fishermen." Importantly, this adaptive capacity of the community was inherently built in as part of the existing local socioeconomic organization, or "life organization" (also see Chapters 2 and 18), represented by a group of fishers who engage in "set net fishing." By extending the logic and practice of the group, originally meant to support local inhabitants, to migrant surfers, the local community has managed to sustain their everyday lives. This chapter therefore spotlights the kind of creativity and ingenuity that local communities tap into in order to resolve pressing problems that inhibit their daily lives without necessarily invoking such normative values as justice and fairness, or such prescribed procedures as democratic consensus-building.

Sustaining Surf Tourism or Sustaining Coastal Communities?

Surfing is commonly portrayed as environmentally benign and culturally progressive (Hill and Abbott 2008). It is true that, in contrast with sports like baseball, soccer, or skiing which require specialized sporting infrastructure, surfing is an activity wherein nature can be utilized without significant alteration. When researchers focus solely on this lack of sporting infrastructure it is possible to portray surfing as an activity with a minimal environmental footprint. However, the relationship between surfing and environmental and sociocultural sustainability is not so simple (Borne and Ponting 2017).

Hill and Abbott (2008) identify four distinct perspectives in political ecological thought that challenge representations of surfing as ecologically and culturally innocuous. The first pertains to surfing's connection to consumer culture. Surfing is commonly represented in media and advertising as an escape from civilization and a chance to reconnect with wild nature – as a network of surfing-related commodities whose consumption can remedy the limited opportunities to commune with "nature" afforded by modern urban life. This wide network of surfing-related commodities is deeply imbricated in modern capitalism and its culture of consumerism. The second perspective points to the environmental burden generated by surfing-related activities (e.g., Gibson and Warren 2017). To meet the needs of surfers, large numbers and varieties of petrochemical-derived products – surfboards, wax, wetsuits – are mass-produced and mass-consumed. In the process of manufacturing such products harmful substances are generated, and large amounts of energy are used both in production and in the consumptive processes of transporting people and materials. The third point concerns the relationship between surfing and neo- or post-colonialism. As a result of surfing-related tourism, coastal areas are transformed into spaces for consumptive recreation – as sites for "exotic surfing" and "endless summers" – and thereby the land and people of these areas become controlled or dominated (e.g., Ponting, McDonald and Wearing 2005). The fourth perspective has been articulated by critical feminists who describe how surfing is constructed both physically and symbolically through male dominance. Hill and Abbott (2008) note, for example, that "the highest compliment a woman surfer can receive ... is to be described as 'surfing like a man'" (291). In this way, they argue that female bodies and assigned gender roles are used as an important component of ensuring male dominance in the culture of surfing.

To be sure, these critical perspectives provide us with useful means to scrutinize the conventional wisdom that uncritically links surfing with environmental and sociocultural sustainability. Nevertheless, existing studies on surfing still tend to fall short in paying close attention to the daily lives of the people who live in surfing areas. Examining surf tourism in Fiji, Ponting and O'Brien (2014), for example, observe the damaging impact of neoliberalization in surf resorts, and advocate for the urgent need to resolve "the seemingly conflicting tensions between the desire for indigenous control over custodial resources and the need for increasingly considerable amounts of tourism investment" (14). This assertion, however, still implicitly accepts the imperatives for tourism development. In a similar vein, Buckley (2002) also expresses concerns for Indonesian island communities threatened by commercial logging and plantation agriculture, but the recommendation is to develop ecological and cultural tourism using surfing as an anchor point while carefully determining recreational capacity and managing local resources. Despite the apparent interest in the well-being of local host communities, the central focus of these two studies is how to make surf tourism sustainable. From this perspective, local host communities are expected to adopt a suitable model of development that accommodates tourism and recreation.

Nevertheless, these models of sustainable tourism are often the creation of researchers and policy makers. As such they risk downplaying the kind of creativity and knowledge/wisdom that are firmly grounded in the lives of local inhabitants and accumulated through their everyday-life experiences. In countries where there is little primordial wilderness and where "nature" has been produced through and in connection with local communities, the creativity and knowledge of transforming nature into the sustainable material basis of human life is accumulated in the life organizations of the local community. From such a perspective, a notion such as "sustainable fishery," for example, does not simply concern the question of sustaining the yield of fish resources, but becomes also about how to secure the human members of the fishery and to maintain the well-being of the community surrounding the fishery.

From the perspective of the life-centered approach, we are compelled to examine surfing in relation to local life organizations. Accordingly, this chapter will analyze the dynamic relation between surfing and the local *Teichi* group in Kamogawa as a key life organization. By doing so, we can begin to understand surfing as part of the everyday-life history of people who live with the sea.

Outline of the Case Study Site

Fishermen and Surfers' Contrasting Views of Kamogawa's Marine Space

Kamogawa City is located in southern Chiba Prefecture along the coast of the Pacific Ocean. In 2020, the city's population totaled 32,116, including 14,578 households. Kamogawa is a roughly two-hour journey from Tokyo by express train, and would be considered outside of the commutable area to Tokyo (Figure 4.1). The current city of Kamogawa has its origins in the fishing village of Ōura, and most fishermen continue to live in Ōura, where the fishing port remains located (Figure 4.2). In the Middle Ages (13th century) the fishing village of Ōura was immediately adjacent to the coast. However, as a result of land being washed away by the erosive action of waves, the Ōura district (pop. approximately 2,500 in 2010) is now located on slightly higher ground away from the coast.

Ōura's fishing industry centers on set net fishing (*teichi ami ryo*, or literally "fixed-in-place net fishing"). This is a "passive" method of fishing in which the

Figure 4.1 Location of Kamogawa. Map by the author.

Figure 4.2 Kamogawa fishing port and Ōura district on the higher ground. Photograph by the author.

only fish caught are those that swim into the fixed-in-place nets, typically set near the coast (Figure 4.3). In Ōura, the term *Teichi* (テイチ) refers both to the method of set net fishing *and* to the fishing group responsible for maintaining this system of nets. The *Teichi* of Ōura was established by willing villagers in 1918 to create a working place for villagers who did not own their own boats. Over the last decade, *Teichi* has had around 30 members, and it is currently managed by the Kamogawa Fisheries Cooperative (*gyogyō kumiai*), of which all fishermen in Ōura are shareholders.

Since the Middle Ages, the fishing village of Ōura has been plagued by strong waves. While there are many memories in the area of hardships caused by large waves, the most famous is the massive wave of April 7, 1951, that hit *Teichi* and killed seven people. A rowboat carrying a large load of yellowtail was capsized by a wave created by a reef zone known as Ichinori (Figure 4.4). As a result of this marine disaster, plans were approved that same year to construct a large and smaller breakwater. Both projects were completed in 1963. This marine disaster is still remembered today as the worst marine accident in the history of Ōura.

Surfing was brought to this fishing town in the early 1960s when U.S. soldiers stationed nearby found Kamogawa to be a great surfing area. Locals recount that children who had previously enjoyed riding waves on washboards began to borrow surfboards from the soldiers to surf. The result was that the eastern side of Kamogawa's fishing port gradually became a surfing point

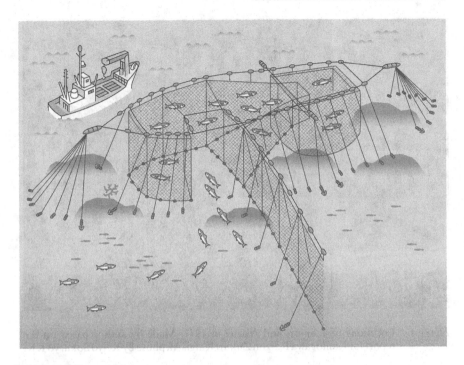

Figure 4.3 Set net fishing.

Source: Ministry of Agriculture, Forestry and Fisheries https://www.maff.go.jp/j/tokei/census/gyocen_illust2.html

known as Akatei ("Red Bank") (Figure 4.4). The reef zone that produces the Akatei surfing point was the same area, called Ichinori, that caused the marine disaster in 1951.

Subsequently, Kamogawa quickly developed as a famous site for surfing in Japan. In July 1964, the first Japan Surfing Competition was held at Akatei. In 1965, the Japan Surfing Federation was established by various surfing groups in Kamogawa. In 1968, Akatei was introduced in the cult classic movie *Endless Summer*, which featured the best surfing points around the world, further popularizing Kamogawa among surfers.

Emergent Conflicts Between Fishermen and Surfers During the 1960–1980s

Since the discovery of Kamogawa as a surfing spot in the early 1960s, broader macroeconomic change in Japan from the late 1960s to 1980s greatly transformed the area, including the Ōura district. First, as many local inhabitants left to find work in the rapidly growing economy of Tokyo, Ōura experienced population decline. Then, to compensate for this loss of population, the fishing industry of the area was rapidly modernized. These changes created the conditions for conflict between fishermen and surfers.

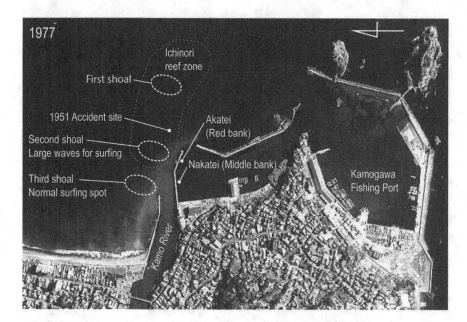

Figure 4.4 Locations of Ichinori and Akatei in 1977. Made by author based on the report for Kamogawa Kaigan Zukuri Kaigi. Airphoto by Keiyo Sokuryō KK.

On March 31, 1967, a new fishing port was opened to the south of the original port to avoid the impact of large waves near Akatei (Figure 4.4). Then, in 1982, *Teichi* replaced its outdated set nets with more durable modern set nets, which were made with synthetic fibers. The background to this modernization of the fishing industry was the loss of population to Tokyo and the subsequent aging of the fishing population. In 1986, the average age of the 38 fishermen on *Teichi* boats was 57.3, with the oldest member reaching 75 years old. Thus, when two 18-year-old youths from Ōura joined and boarded the *Teichi* boats in 1986 and 1987, the people of Ōura were delighted and held high hopes for these two young people. The *Teichi* group even sent them to "study abroad" for three and a half years in the Hokuriku region, which has advanced techniques for set net fishing, evidencing the acute sense of crisis among *Teichi* about the eroding base of fishing.

There were other manifestations of the sense of crisis fueled by aging and a lack of successors in the fishing industry; most notably, the use of national-level planning strategies to promote local tourism development. Based on the national-level "resort law" that aimed to promote tourism development throughout Japan, the Kamogawa Marina Development Project (KMDP) was initiated in 1987. Although the project was framed as tourism development, the purpose of the KMDP was actually to ensure the safety of fishing operations and stabilize fresh fish shipments by controlling large waves. More specifically, the marina, called Kamogawa Fisherina, and a new access bridge were constructed to the northwest of the old port, which led to the destruction of the Ichinori reef zone that had plagued the area with potent waves (Figure 4.5). Even though the project was carried out under the pretense of tourism development, the marina project was actually a desperate measure by the fishing community of Kamogawa to adapt state-led tourism development imperatives to modernize and upgrade the fishing industry. From surfers' point of view, the greatest consequence of the destruction of the Ichinori reef by the KMDP, if completed, would be the disappearance of the surf point Akatei, one of the birthplaces of Japanese surfing.

In late 1980s, citizens' movements protesting against tourism projects by large Tokyo-based developers began to become highly active. These citizens began to see the downside of tourism development in their everyday lives, including the destruction of the local natural environment, pressure on waste management, increased illegal parking, littering, and noise pollution. The KMDP became a symbol of these issues. The movement against the KMDP was initiated by local women who often felt the impact of the project most directly in their everyday lives and through their household work. Some of the problems included increased noise and littering, cinders left after fireworks and outdoor cooking in children's playgrounds, and increased prices. However, as the opposition movement gained steam, Ms. Tachikawa (pseudonym), a surfer who had migrated to Kamogawa in the late 1970s, became the key advocate for the movement.

Figure 4.5 Kamogawa Fishing Port after the development of the Kamogawa Marina, January 2003. Made by author based on the report for Kamogawa Kaigan Zukuri Kaigi. Photo: Keiyo Sokuryō KK.

The movement eventually grew into a highly contested political showdown. On July 8, 1990, a mayoral election was held in Kamogawa which pitted Ms. Tachikawa – running on the platform that tourism development would destroy nature and livelihood – against the head of the fishery cooperative – running on the platform that without decisive action the fishing industry would disappear and Kamogawa would become a tourism-only town. The result of the election was a decisive victory for the head of the fishery cooperative. This contested election remains a vivid memory for the citizens of Kamogawa because this was the only mayoral election in the city between 1987 and 2009; six other elections during the 22-year period had a single candidate each time. Following the result of the mayoral election, the marina construction began in February 1990 as planned, and was completed in June of 2006, resulting in the disappearance of Akatei.

It is important to note that the Akatei issue, which had broiled over into a hotly contested mayoral election, cannot be reduced simply to the matter of marina construction for tourism development. In fact, fishermen, surfers, and local women were all trying to cope with the tremendous external pressure forced on them by tourism development promoted by Tokyo-based capital. However, for each actor their perceived problem and preferred resolution were quite different, depending on the very different ways in which each actor experienced the waves and the sea. To put this more concretely, the spatial alteration of the Ichinori reef zone was viewed by fishermen as a way to control the waves that had plagued them for years, and by surfers as part of the overdevelopment leading to the disappearance of the surfing point Akatei which they cherished.

Endless Conflict

The mayoral election and the ensuing marina construction did not end the controversy; rather, surfers and fishermen began to organize and intensify their efforts to further pursue their goals. In 1993, surfers, led by Ms. Tachikawa, held a rally against the proposed breakwater next to Kamogawa Fisherina that was being built (Figure 4.5). Hundreds of surfers gathered from all over Japan to protest against the breakwater project at the rally. Their zeal quickly refracted against the marina under construction, and surfers straddling their surfboards protested from the sea. This protest drew attention from the Surfrider Foundation, an organization founded in 1984 in the United States that aims to protect marine and adjacent environments from the perspective of surfers and bodyboarders. In 1995, a Surfrider Foundation Japan (hereinafter SFJ) branch was founded in Kamogawa and Ms. Tachikawa was appointed as its CEO.

Teichi fishermen, on the other hand, sought further to modernize their work. Mr. Sakagami, one of the young fishermen who had "studied abroad" in Hokuriku, returned to Kamogawa in 1993, and became the captain of *Teichi* at the age of 28. As a new captain, Sakagami, with the slogan of "new challenges," aggressively implemented various measures to strengthen their industry with particular foci on preventing damage to fishing equipment, increasing the efficiency of work, and increasing the value-added of catches (Sakamoto 2009).

As discussed further below, *Teichi* also began to accept in-migrant surfers as members in 1998, resulting in a steady shift of the generational guard (Figure 4.6). Their interest was squarely to sustain the set net fishing in Ōura.

An interesting attempt to bridge the gap between surfers and fishermen took place on November 11, 2003, when, through the efforts of SFJ, the first Annual Coastal Planning Conference was held by Kamogawa City. This conference brought together experts, fishermen, surfers, and tourism businesses to democratically deliberate on the future direction and use of coastal and marine areas. However, most of the participants other than the organizers had a rather chilling opinion such as "they had done nothing more than try the American way in Japan" (personal interview with Ms. Tachikawa). "The American way" here assumedly implies something along the line of consensus formation through a fair and open debate based on logic and reasoning (akin to the idea of procedural justice). A total of six meetings were held between 2003 and 2007, and surfers submitted proposals based on their reflections on the Akatei issue and proposed for the removal of the breakwater. What became clear, however, was that the more meetings they held, the more apparent the difference in discursive logic between stakeholders became, rather than reaching some sort of mutual agreement. Eventually, the conference was disbanded and meetings ceased to be held.

Seeing the city-level meeting heading nowhere, SFJ held the SFJ International Conference in Kamogawa from October 8 to 12, 2010. Under the theme of "Coastal Environment Conservation Activities and Exploration of New Uses of Coastal Areas," prominent surfers and environmental protection activists gathered from around the world to introduce coastal management methods from beyond Japan, investigate Japanese coastal management methods, and exchange

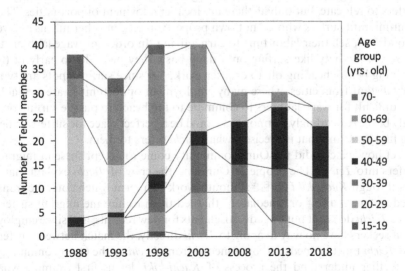

Figure 4.6 Age compositions of *Teichi* members.
Source: Author's survey.

opinions. In addition, a movie titled *The History of Akatei* was screened, in which the overarching message was to remember Akatei, along with its role in the history of surfing in Japan, as "the past that must be reflected upon."

In a small city like Kamogawa, the series of activities centered on SFJ since 2000 is a widely known story among the residents. Indeed, that was precisely the goal of surfers who sought such attention. The activities of the surfers centered on the SFJ can be understood as citizen participation wherein new users become actively involved in coastal and marine management, activities which had previously been exclusively conducted by the government and fisheries cooperative. Such active civic participation may be considered increasingly important and desirable in contemporary environmental governance. However, the type of civic participation and mutual engagement seen in the case of Kamogawa has accentuated the divide between surfers and fishermen.

Teichi as a Life Organization: The Logic of *Kuchi-kiki* (Mediation)

The views of fishermen and surfers remain sharply divided; that is, they are in a situation of mutual non-understanding (see Chapters 2 and 18). Yet, the Ōura community appears to accept and embrace surfers, often as *Teichi* members, in their everyday lives. In 1998, *Teichi* leaders began hiring in-migrant surfers and, as of 2019, about 60% (16 of 27) of members of the *Teichi* crew are in-migrants. Of those 16 in-migrants, 11 are surfers who came to Kamogawa for its waves. This brings us to the question of why *Teichi* has become a key receptacle for in-migrant surfers, and compels us to pay closer attention to the specific functions of *Teichi* as a local life organization.

Even though the Akatei surfing point no longer exists, Kamogawa still has a handful of great surfing points, making it a highly attractive community for surfers to relocate, but only if there are local employment opportunities. That is, in-migrant surfers who do not own property, boats, or other means of livelihood must sell their labor time to earn a living. In order for wage laborers to pursue an activity like surfing on a daily basis they must do so early in the morning before heading off to or after work. Yet since surfing spots are typically distant from cities, where many employment opportunities are available, it is difficult for wage laborers to commute to the beach to pursue surfing on a daily basis. Accordingly, Kamogawa would be a perfect place for surfers if they can find jobs, and that is precisely what *Teichi* offers to them.

Yet how exactly did the Ōura community come to accept these in-migrant surfers into *Teichi*? The people of Ōura and the crew of *Teichi* explain that it was through *Kuchi-kiki* (口利き: "mouth-work"), a term which could be translated as "mediation" or "mediator," that led them to hire the migrant surfers. Since *Teichi* does not put out advertisements for new recruits, and since employment seekers have no way to apply to *Teichi* directly, the match between surfers and *Teichi* must be mediated by someone from *Teichi* or the Ōura community. To further understand the process of *Kuchi-kiki*, let us first examine what *Teichi* means to the Ōura community.

Teichi *as a Life Organization*

The significance of *Teichi* for the people of Ōura as a whole can be deciphered from the fact that *Teichi* is an indispensable topic within their daily conversations. They often refer, in their local Bōshū dialect, to the set nets as "*Oraga* (meaning 'our/my') Teichi," a phrasing which clearly indicates a sense of collective ownership. The reason for this sense of collective ownership is that *Teichi* is not simply a fishery organization that serves a productive function. It is, rather, a life organization that enables the local community to maintain and adapt their communal lives in the face of an ever-changing external environment. We can summarize the three functions of *Teichi* that enable such adaptation.

As a Social Safety Net

In recent years, an increasing number of young people have moved to Kamogawa to flee urban areas in search of waves or, more generally, a nonurban pace of life. These young people are migrants pushed by macro-scale sociocultural changes. *Teichi*, which was originally created as a place of employment for the inhabitants of Ōura, has become a sort of social safety net for these young in-migrants. In the past, whenever the people of Ōura have found themselves in need of employment – for example, when many jobs were lost due to the decline of pelagic fisheries after the setting of the 200 nautical mile fishery limit in 1977, when former residents returned from urban areas, or when senior citizens are looking for employment after retirement – *Teichi* has offered them a place to earn a living. The acceptance in recent years of surfers and migrants from urban areas into *Teichi* should be seen as a continuation and further extension of this long-standing social safety net function of *Teichi*.

As a Means to Respond to Fickle Market Conditions

Teichi supports not only the economic livelihood of individual households, but more broadly the economic base of Ōura as a whole, keeping the community resilient in the face of ever-fluctuating market conditions. For example, *Teichi* provides proportionally more funds to the Kamogawa Fisheries Cooperative. The port usage fees paid to the cooperative are 10% of the value of landings (catch) for the community-owned *Teichi*, while this figure is only 4% for other fishing groups. In addition, when annual sales from *Teichi* exceed 500 million yen, dividends are paid to all of the fishing households, which comprise its shareholders. Informants recall fondly that group trips with the dividends from *Teichi* always constitute a fun topic of conversation at gatherings.

In contrast to "active" fishing methods that target specific fish, a "passive" method such as set net fishing lands a wide variety of fish species. While it may be impossible to haul in a tremendously lucrative catch through set net fishing, this method does provide a stable supply of seasonal fish to the market. For this reason, fish wholesalers from throughout the Minamibōsō Peninsula flock to Kamogawa Port, leading prices at the port to stabilize at higher levels. Many people also come to the port because it is one of the few ports where you can frequently find stable supplies of mackerel and other rare fish. Currently in Ōura even fishermen with aging equipment and bodies and with smaller catches of less-valued fish species are able to earn a decent income.

As a Means of Managing Everyday Community Functions

Teichi's regular activities involve far more than just fishing. For example, port maintenance and water rescue are also carried out under the organizational umbrella of *Teichi*. Even more unique functions of *Teichi* include care and maintenance of local religious institutions and infrastructure. These include Yagumo Shrine (the resident deity of Ōura), Batōkannon (Hayagriva: the god of maritime safety), Yamazumi-sama (the god of abundant fishing), as well as Daikoku-kō (a group that worships the god of luck), and Tenjin-kō (a group that worships the god of learning). While care of these religious institutions was originally carried out by parishioners (*ujiko*), due to the aging of the area's population and decreased birth rate, this has become quite difficult. As a result, *Teichi* has now become the central agent in maintaining religious rituals, many of which are important symbols of the communal life of Ōura (Figure 4.7). For the people of Ōura, then, *Teichi* is not only a means for dealing with social and economic changes from outside, it is also a life organization for dealing with various problems that arise within the community.

Through *Teichi*, the people of Ōura have continued to adapt their daily lives in response to major socioeconomic changes – from national-level policies to markets – and also to issues arising within the community. If *Teichi* were to become a standard fishery organization that pursues profits and becomes separate from the lives of the people of Ōura, it would no longer be *Teichi*. For that reason, the head of *Teichi* takes on the responsibility of ensuring that this life organization will continue to serve the Ōura community as a whole. Accordingly, if someone from Ōura asks *Teichi* for a favor of *Kuchi-kiki* (mediation), such as hiring a relative of the Ōura resident, the head will not and cannot turn down the favor. Indeed, no one has ever been "fired" from *Teichi*.

Extending Kuchi-kiki *(mediation)*

Despite the continuing importance of *Teichi*, after it began hiring surfers in 1998, the meaning and working of *Kuchi-kiki* has gradually changed. Previously, both the person who sought a go-between and the individual who served as an intermediary needed to be somehow connected to Ōura or *Teichi*. However, since the 1990s, Kamogawa has become an in-migrant city that receives a 5% population influx annually. Everyday social relations in this in-migrant city are formed regardless of the distinctions between locals ("*tochimon*") or nonlocals ("*yosomon*"). For example, Mr. Tokuda, a man who came to Kamogawa as a high school student for a surf study tour, married a young woman from the area and was able to join *Teichi* through the intermediation of his father-in-law.

Figure 4.7 Migrant surfers, also members of *Teichi*, carrying the local deity of Ōura. Photograph by the author.

Mr. Ishiyama, who was having difficulties due to being unemployed, joined *Teichi* through the intermediation of a surfing friend from Ōura. When Mr. Noda was having difficulty due to an unstable job, he was introduced to *Teichi* through the intermediation of a "drinking buddy."

In these ways, the distinction between local and nonlocal no longer matters, and both can be afforded *Kuchi-kiki*. Even if the individual in question is an "outsider," if the person is introduced as a good person who is a "friend of a friend," then the collective sense of "Our Teichi" will not and cannot refuse the request. In sum, the acceptance of migrant surfers into *Teichi* is a logical extension of *Teichi*'s role of providing protection for the Ōura community's well-being as a local life organization that has been cultivated across generations. While its first priority is to continue serving as a local life organization for Ōura, it has also been extended to serve as a life organization that safeguards the daily lives of in-migrants as well.

I wish to call this relationship between fishermen and surfers, mediated through the *Teichi* organization and *Kuchi-kiki* practice, coexistence at the level of daily life. It is not a relationship that has been attained through mutual understanding based on "fair and open" debates and democratic delivery, as was attempted by the city government in the early 2000s. Rather, opposed parties temporarily shelved points of contention around the use and meaning of waves and oceans, in favor of prioritizing each other's daily lives and livelihoods. Extending the indigenous logic of *Kuchi-kiki* to the nonlocal, migrant surfers is what enabled such everyday coexistence.

Conclusions

Due to its warm seaside climate and relative proximity to the Tokyo metropolis, Kamogawa has experienced rapid tourism development as well as large population inflows and outflows. Even though fishermen and surfers were both concerned about the rapid touristification of Kamogawa, the supposed "solution" – the proposed marina construction – resulted in sharply divided responses. This is because wave-hating fishermen and wave-loving surfers had completely different experience and understanding of the ocean space, leading to markedly different ideas of what ought to be done.

As Satō (2014) states, the natural environment has no value in and of itself; rather, the value of nature is determined through its historically formed relation to society, and according to the place-specific social relations in which nature entangled. Even though the series of conferences organized by the city and SFJ in the early 2000s were organized with the aim of democratic consensus formation, this venue made the difference in their views more pronounced. This is because the forum overlooked the historically formed and geographically specific socio-environmental relations, and intentionally created stakeholder groups based on participants' social attributes and positions. In contrast, the people of Ōura tried to reposition the surfers as "friends of friends" within the norms of Ōura by using *Kuchi-kiki* as an "everyday-life-enabling technique." Rather than applying the social category of "surfer" to these in-migrants, the people of Ōura reconstructed their relationship with them by seeing them as fellow inhabitants.

The acceptance of surfers into *Teichi* did not happen because the people of Ōura came to thoroughly understand the logic of surfers. It occurred instead through the extension of the already-existing logic of "Our Teichi" and the associated practice of *Kuchi-kiki*. It would be easy to criticize *Kuchi-kiki* as a form of nepotism, far from being fair, transparent, or democratic. However, inquiries that start out with such principles and ideals, abstracting away historically formed unique and concrete details of the people and place, risk overlooking the life challenges local inhabitants actually face as well as the creativity they deploy to overcome them. The story of Ōura shows that even if they do not completely understand each other, the adaptation of existing life organizations, such as *Teichi*, may enable their coexistence at the level of daily life.

References

Buckley, Ralf. 2002. "Surf Tourism and Sustainable Development in Indo-Pacific Islands. I. The Industry and the Islands." *Journal of Sustainable Tourism* 10 (5): 405–24. https://doi.org/10.1080/09669580208667176

Borne, Gregory, and Jess Ponting. 2017. *Sustainable Surfing*. Oxford: Routledge.
Gibson, Chris, and Andrew Warren. 2017. "Surfboard Making and Environmental Sustainability: New Materials and Regulations, Subcultural Norms and Economic Constraints." In *Sustainable Surfing*, 87–103, edited by Gregory Borne and Jess Ponting. Oxford: Routledge.
Hill, Lauren L, and J Anthony Abbott. 2008. "Surfacing Tension: Toward a Political Ecological Critique of Surfing Representations." *Geography Compass* 3 (1): 275–96. https://doi.org/10.1111/j.1749-8198.2008.00192.x
Ponting, Jess, Matthew McDonald, and Stephen Wearing. 2005. "De-Constructing Wonderland: Surfing Tourism in the Mentawai Islands, Indonesia." *Loisir et Société / Society and Leisure* 28 (1): 141–62. https://doi.org/10.1080/07053436.2005.10707674
Ponting, Jess, and Danny O'Brien. 2014. "Liberalizing Nirvana: An Analysis of the Consequences of Common Pool Resource Deregulation for the Sustainability of Fiji's Surf Tourism Industry." *Journal of Sustainable Tourism* 22 (3): 384–402. https://doi.org/10.1080/09669582.2013.819879
Sakamoto, Toshikazu. 2009. "Kotsukotsu Tsumiageta Teichiami Gyogyō no Gijutsu" [The Steady Improvement of Stationary Net Fishing Technology]. *Zenkoku Seinen Josei Gyogyōsha Kōryū Taikai Shiryōi* [National Youth and Female Fisheries Exchange Conference Materials] https://object-storage.tyo1.conoha.io/v1/nc_a1d807edab8b4dde9d9e321cea76c59c/jf/6f7ded4ac1e2b7e63a70c0813238a6b7.pdf
Satō, Jin. 2014. "Shizen no Shihai wa Ikani Ningen no Shihai e to Tenzuru ka–Komonzu no Seijigaku Josetsu" [How Domination of Nature Turns to Domination of Humans: An Introduction to Political Science of the Commons]. *Nihon no Komonzu Shisō* [Japanese Common Thought], edited by T. Akimichi, 176–94. Tokyo: Iwanami Shoten.

5 When Civil Society Falls Short
Rural Community Response to a Resort Development Project*

Daisaku Yamamoto and Yumiko Yamamoto

The postwar Japanese developmental state was characterized by its mechanism of "developmental redistribution," which aimed to counter growing interregional economic disparities by implementing more growth projects in rural peripheries. These projects took various forms, such as large-scale energy infrastructure, industrial parks, and leisure resorts, all under the banner of "helping backward regions" and promoting national economic growth. These developmental projects strike as external "shocks" to rural communities. Social scientists have often focused on studying how local communities react to such developmental projects.

This chapter examines the historical experience of a rural Japanese village community that responded to a golf resort project brought by a Tokyo-based developer company during the 1970s.[1] The community's response was complex, but the most visible was a grassroots environmental movement against the project. After five years of struggle, the project ultimately stalled, primarily due to the developer company's bankruptcy but also due to significant local resistance that prevented a swift realization of the project prior to the bankruptcy. At first glance, this case appears to be a story of a burgeoning modern civil society and local activism in a rural periphery, led by a talented local leader who had connections to external resources. In other words, it seems like a classic example of a successful civic movement characterized by strong social capital and participatory democracy that effectively resisted an unwanted land use project in the community.

However, the reality of the conflict was far from simple. Through the analysis of the golf resort project controversy, this chapter reveals that the "civil society interpretation" provides only a partial understanding of the whole story. It particularly highlights the significance of long-standing indigenous institutions that provided a less-than-democratic basis for resistance and resilience. The chapter also explores the role of "local outsiders," recent settlers who, despite being newcomers, had strong ties to the land. Additionally, it

---

* Empirical sections of this chapter were originally published in Yamamoto, Daisaku, and Yumiko Yamamoto. 2013. "Community Resilience to a Developmental Shock: A Case Study of a Rural Village in Nagano, Japan." *Resilience* 1 (2): 99–115, and are used with permission.

DOI: 10.4324/9781003185031-6

emphasizes the importance of expedient uses of "everyday knowledge," mobilized by the local residents. The chapter begins by providing theoretical contexts for the case study, drawing on insights from life-environmentalism research, and then presents an empirical account of the "failed" golf resort project in a rural village in Japan.

Theoretical Contexts

When analyzing the response of local communities to externally originated projects, observers typically focus on the community's "resistance." While understanding community resistance is undoubtedly important, we should also consider the social resilience of a community, which refers to how a community recovers and maintains a sense of "normalcy" in their everyday lives despite the distress caused by a project. It is rather unusual for these projects to be ubiquitously opposed by all local residents; instead, they often lead to different opinions and tangible impacts within the community.

For those living in local communities, "successful" resistance against a developmental project often does not mark the end of the story. Regardless of the material outcome of the project (e.g., whether a golf course is ultimately built or not), it can lead to serious social divisions with the community. Numerous journalistic and scholarly accounts of golf and ski resort projects in rural peripheries of Japan have highlighted the long-lasting social fissures caused by such projects. The negotiation and decision-making processes associated with these projects tend to generate corruption and a distrust of politics, suspicions of nepotism and favoritism, and antagonism among different segments of the local population (e.g., landholders and non-landholders).

Under such conditions, unlike external actors who can choose to exit or remain only partially attached to the place, most local residents wish to continue living in their communities regardless of the outcome of the controversy. Therefore, the ability of a community to maintain its "normal" everyday life during and after a divisive issue becomes a critical concern.

If both community resistance and resilience are important, the next question is: what factors are necessary for a local community to demonstrate strong resistance and resilience? According to Wilson (2012, 23), key qualities that enhance community resilience include strong social ties, well-established trust, and participatory, inclusive, and democratic processes. In this regard, civil society groups and local environmental NGOs are often seen as critical actors promoting these qualities. Some observers go as far as claiming that they are the key to the future of environmental governance in Asia (Zarsky and Tay 2000; Huang et al. 2020).

Nevertheless, Kuan-Hsing Chen, a Taiwanese cultural theorist, asserts that "civil society" is a relatively new concept in Asia, and that "translated notions such as civil society and citizenship were not smoothly absorbed into the political culture of many spaces in Asia as they underwent modernization" (Chen 2010, 237). Chen points to the existence of *"minjian"* (literally "people-space" or roughly translated as "folk society") as a space where traditions are preserved and serve as resources to help common people navigate the disruptions caused by the modernizing state and civil society. Chen contends that *minjian* did not disappear during the process of modernization but rather represents an evolving set of practices that are increasingly visible in the present (241). Therefore, our study focuses on the intersection between civil society and *minjian*, exploring how these two concepts interact and influence community dynamics.

A case in point is the *jichikai* system, a type of neighborhood community association in Japan. The *jichikai*, which translates to "self-governing association," is known by different names in different regions, such as *chōnaikai*, *burakukai*, and *ku*. In rural parts of the country, the boundaries of *jichikai* often loosely resemble those of "premodern" villages (*mura*) that existed before the Meiji Restoration in 1868. Although these villages no longer exist officially, they continue to influence the mind and behavior of rural inhabitants in Japan. Torigoe (1994, 1997) summarizes the distinct characteristics of the *jichikai* as follows: (1) it has clear geographical boundaries; (2) households, rather than individuals, form the basic unit of the association; (3) membership is near-mandatory; and (4) it is considered a legitimate representative of the community.

The *jichikai* is often seen as a cultural legacy of Japanese society, and is something to be replaced by more desirable forms of governance such as the "new public" or modern civil society. Yet, if we consider Chen's argument, the *jichikai* has likely co-evolved with the emerging forms of governance, such as local environmental NGOs and other voluntary associations.

Lastly, our case study also raises a question about what constitutes critical "knowledge" in socio-environmental conflicts. In this regard, we draw on the notion of "everyday knowledge" as discussed by cultural anthropologist Motoji Matsuda.[2] According to Matsuda (2013a, b), when studying local socio-environmental conflicts, social scientists often analyze the competing discourses that circulate among the locals. These discourses may emphasize the importance of some "value standards" such as the respect for human rights, the need to break class barriers, or the urgency of protecting tribal culture. However, Matsuda (2013a, b) argues that the essence of the life-centered approach (*seikatsu-ron*) is not to place excessive trust in the presence of such value standards or to view the discourses as intense battles between values or ideologies.

This is because the local inhabitants are typically very aware of the origin of discourse and how they are rooted in the concrete local experiences of the community. For example, when Resident X objects to the building of a resort project, stating that it is a neoliberal capitalist exploitation, other residents know that Resident X is primarily concerned with protecting his farmland. In such circumstances, it is inadequate, or even misleading, to read into the essential meaning, or to analyze logical coherence of the discourse itself. Matsuda argues that there is usually a stock of value standards and associated discourses (or *iibun*) that justify the standards in the consciousness of the people living in a local community. These inhabitants choose a particular discourse "out of convenience and the circumstances of their lives" (Matsuda 2013b, 26), rather than through a careful and thorough reflection of their action and consistency with their worldview and value standards. Matsuda emphasizes the expedient aspect of everyday knowledge by using the specific term *bengi-chi* ("expedient knowledge") in other writings (Furukawa and Matsuda 2003). This term conveys more explicitly that it refers to meta-knowledge that guides the selection and application of specific knowledge suitable for particular circumstances.[3]

For Matsuda, everyday knowledge serves as a means for individuals to select a particular discourse, reasoning, or value in a given situation, thereby supporting the practical functionality and continuity of everyday life. In our view, the concept of everyday knowledge differs in emphasis from such notions as gendered knowledge, indigenous knowledge, or traditional knowledge, which pertain more prominently to alternative systems of knowledge (Elmhirst 2015, Mazzocchi 2020; Gadgil, Berkes, and Folke 1993; Bruchac 2014).

Matsuda (1989) argues that everyday knowledge assists local communities in dealing with external forces of domination, sometimes by seeking compromises, temporarily submitting, or even resorting to violent resistance. Therefore, in considering both the resistance and resilience of a community during socio-environmental crises, everyday knowledge may play a crucial role. Our study sheds light on the nuanced yet significant role of everyday knowledge in the local golf resort controversy.

Matsukawa Village: Geographical and Historical Settings

Matsukawa Village, the case study site, is located in the Azumino basin at the foothills of the Northern Japan Alps. The majority of the village's 10,000 residents reside on the eastern side, offering picturesque views of snow-capped mountains that persist into early summer. The western half of the village is mountainous and forested (Figure 5.1) while the Ashima River has shaped a gently sloping alluvial fan, overlooking rice paddies intermingled with scattered houses to the east (Figure 5.2). The golf resort project was planned for the forested alluvial fan area known as Gōdohara.

From the mid-1950s through the late 1960s Japan experienced rapid economic growth fueled by industrialization along the Pacific Manufacturing Belt. During this period, a significant migration of young labors from rural to urban areas resulted in rural decline, which became a major political issue. In an effort to counter this trend, "mini-growth" projects were initiated in disadvantaged regions, often under the mantra of "alleviating regional disparities." One such project was the Act on New Industrial Cities (1962), a government-sponsored initiative aimed at promoting manufacturing industries in selected "local core cities" outside the Pacific Manufacturing Belt. For more remote rural areas with limited prospects for attracting major manufacturing activities, the promotion of leisure and tourism for urbanites emerged as a more viable option. Reflecting the trend, the number of golf courses in the country, for example, increased from 74 in 1957 to 263 by 1961.

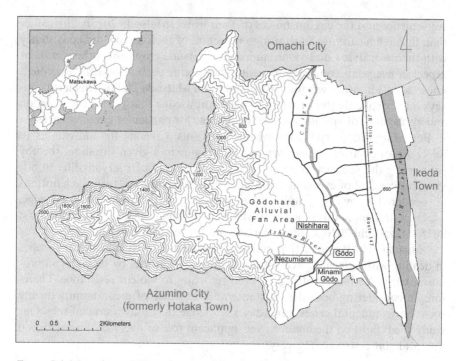

Figure 5.1 Matsukawa Village, Nagano. Map by the author.

Figure 5.2 Gōdohara alluvial fan area on the outskirts of the Japan Alps (Hida Mountains). Photograph by the author.

The village of Matsukawa was among the rural municipalities that sought to benefit from these regional development initiatives. In 1963, the Matsumoto-Suwa region of Nagano Prefecture was designated as one of the New Industrial Cities by the central government. Located on the fringe of this designated area, the village was eyeing on diversifying its rice-dominated economy by promoting leisure and tourism. Matsukawa's southern neighbor, Hotaka Town, had already opened a golf course in 1972. Given these circumstances, the Gōdohara alluvial fan area, which was no longer used for small-scale timbering and firewood collection, was considered as potential land for development.

Gōdohara Golf Resort Project

In December 1972, Sankyō Kaihatsu Co., based in Tokyo, presented a proposal to develop an 18-hole golf course and five nature park zones in the Gōdohara area. Inconspicuously, Sankyō had already 61 hectares of the land through regional real estate agents. After the turn of the year, the mayor of Matsukawa, Ichiro Takada, expressed an approving stance on the project. Members from key local organizations, including the village council, the committee on agriculture, the chamber of commerce, and the agricultural cooperatives, formed the Commission on Natural Environment Conservation (CNEC). After visits to golf courses in nearby municipalities, meetings with Sankyō, and consultations with other relevant government offices, the CNEC submitted a list of recommendations to the village office. The list included guidelines for forest conservation (preserving at least 40% of the land), drainage treatment, flood prevention, and road construction/maintenance. Sankyō modified the golf project based on these recommendations, and it was subsequently accepted.

Following this acceptance, the village office followed a normal protocol of consulting with "affected districts," referring to the four districts (i.e., *jichikai*) adjacent to the proposed site of the golf resort. This consultation is a convention that is still followed scrupulously whenever important village-level decisions need to be made. This practice grants a relatively strong autonomous power to adjacent *jichikai*, thereby adding another layer of governance alongside the village council with representatives, which operates as a more formal democratic decision-making system. During these consultations, two of the four "affected districts," namely, Nishihara District and Gōdo Districts, opposed the project, while the Nezumiana District and the Minami-Gōdo District supported it. As a result, the project was not about to move forward quickly.

Civil Society Organization: Its Success and Limitations

While the village office was contemplating the project, a local movement opposing the golf course project emerged. Led by Takashi Tada (deceased; pseudonym), a mountaineer and native of Matsukawa, the Matsukawa Nature

Conservation Society (MNCS), a civil society organization, was established on January 11, 1974. Tada, confident and charming, successfully rallied dedicated friends and followers to advance the movement. The MNCS worked closely with the residents of the Nishihara District, while Tada used his personal connections to invite experts to highlight adverse effects of the project. The mobilized scientific and cultural knowledge encompassed a wide range of concerns, including the increased risks of groundwater pollution, floods, landslides, the value of ecological preservation, and the cultural heritage of the area (e.g., the significance of sacred rock formations).

In less than a month, the MNCS collected 2,919 signatures opposing the golf course project, a clear majority of eligible voters in the village of 6,636 people at that time. On January 31, after receiving the petition from the MNCS, the mayor made the decision that the village would not permit the golf resort development, noting that the majority of the local residents and two affected districts were opposed to the project. The civic movement achieved a significant victory in a short period, but this was just the beginning of the story.

As of September 1974, the mayor's attitude appeared unchanged, as indicated by his comment during the regular council meeting,

> I think that it would be very difficult for the village to promote tourism, considering the geographic and climatic conditions of our village. Therefore, we have no plan to promote tourism at this point. Even if private companies propose tourism development, we will not go forward with it unless the village residents approve it.

However, around this time, Sankyō took an offensive approach and began conducting negotiations behind the screen with the affected districts. They offered "cooperation money" and preferred employment opportunities for their residents, while also engaging in lobbying efforts targeting village officials and council members.

On March 16, 1975, 13 months after the mayor had declared the project dead, Sankyō submitted a memorandum to the mayor and the local council head, expressing their renewed interest in developing a golf resort. The memorandum emphasized the company's increased commitment to environmental protection, which included: (1) ensuring water supply from the company's deep well in case of water depletion in surrounding districts, (2) providing compensation for any damage to agricultural products caused by drainage, and (3) canceling the cottage development near the golf course.

During the interpellation session of the regular council meeting in April 1975, several council members expressed a favorable view of the golf course development. Sensing the shift in sentiment, the MNCS exerted pressure on the mayor to maintain his opposition. However, the approval process for the golf course project gained momentum when the Nishihara District changed its stance and issued a "conditional approval" for the project on May 27, 1975. The unexpected change in the Nishihara District's position requires further explanation, but for now, we focus on the actions of the local council. On June 2, the mayor officially declared that all affected districts had now approved the proposal, and the village was prepared to authorize the development of the golf resort (by this time, the Gōdo District had also given its approval).

Following the mayor's announcement, the local council swiftly took action, which seemed deceitful in the eyes of those who were against the project. The council formed the Special Committee on the Golf Course, conducted another round of hearings involving Sankyō and the MNCS, visited several nearby golf courses, and the committee ultimately approved the project on the same night. Based on this decision, the project was approved, with a vote of 13 to 1, at the extraordinary council meeting on August 12, 1975. A new series of consultations with the affected districts was conducted in order to finalize the agreement.

The MNCS did not remain idle after the mayor and the local council shifted toward approving the project. In July 1975, the MNCS submitted a letter of complaint to the village government, and in August, it held an emergency meeting following the Special Committee's decision to approve the project. Even after the final approval by the local council, the MNCS persisted in expressing concerns about the potential adverse effects of the golf course development and continued its public campaign. However, it became clear that the MNCS did not have direct leverage to control the fate of the project at this stage.

The MNCS undoubtedly played a critical role in this controversy; it raised awareness of the project among village residents and had a visible impact on the mayor's decision to temporarily hold back from the project. The charismatic Tada played a crucial role in mobilizing supporters, utilizing external resources such as professors from Shinshu University, and articulating concerns about potential risks using modern scientific language. In many ways, the MNCS was an exemplary modern civil society organization, and without their involvement, the construction of the golf resort might have started before Sankyō filed for bankruptcy.

However, the experience of the MNCS also highlights potential limitations of modern civil society organizations. First, it was challenging to sustain active participation from residents over an extended period. This was especially true for residents who did not live near the project site, as the project's impact on their everyday lives was less apparent. After the initial enthusiasm, many local residents lost momentum. Second, as a consequence, external agents such as the media and the broader public also lost their initial interest and active support. Third, unlike *jichikai* (neighborhood associations), voluntary organizations like the MNCS were not considered legitimate representatives of the residents and faced the constant risk of being dismissed as "mere activists." These risks may still be present in contemporary civic movements. Our focus then shifts to the roles played by the "old" institution of *jichikai* alongside the MNCS. Specifically, we examine next the Nishihara District, which maintained close relations with the MNCS throughout the controversy.

A Contemporary Role of Premodern Local Institution

The Nishihara District is located at the tip of the Gōdohara alluvial fan area, directly below the proposed golf resort site. One might imagine that the district would oppose the project. However, as mentioned above, the district shifted from opposition to approval, albeit with certain conditions. What prompted this change in the Nishihara District's position? Furthermore, the role of Nishihara raises a question about how the *jichikai* influenced the course of this controversy. To answer these questions, it is first necessary to look more closely at the district's history.

In 1945 as part of the country's postwar reclamation project to increase food production and to defuse potential social discontent among the young population, laborers aged 15–19 were sent to Matsukawa Village and other rural areas of the country to convert the land. The Gōdohara area was originally forested with red pine trees and filled with granite boulders, but it was transformed into wet rice paddies. By 1972, when the Reclaimed Land Agricultural Coop of the village was progressively dissolved, the District had 42 households of a uniquely young demographic profile. Very few people were over 60 at the time of the golf resort controversy. In the eyes of long-term village residents, these young settlers were considered "outsiders," further strengthening bonds among the young people. The district also had a relatively large number of Japanese Communist Party members. Most importantly, many of these young residents were second or third sons in their families, which meant they had no defined role in their hometowns where the first son traditionally inherited the household. For the young residents of Nishihara, making a living from the land in Matsukawa was their only viable option, creating a paradoxical situation where they were both locally bound and outsiders. Their strong opposition to the golf course cannot be fully understood without appreciating their position within the village community and their deep emotional attachment to the land.

Although we were unable to find any written records explaining why the residents of Nishihara District reversed their earlier opposition to the golf course on May 27, 1975, interviews with several long-term residents revealed the story. It began with a change in the Nishihara District's chairperson. As per convention, the district rotated the position of chairperson among the heads of households on a regular basis. The new chairperson was a moderate individual who was not a natural leader. He was not particularly familiar with the controversy surrounding the golf resort, as he was not part of the special golf resort project commission that had been working closely with Tada and the MNCS.

Sometime in April or May, a village official visited the new chairperson at home and persuaded him to sign a document without reading it. The document announced the district's approval of the golf resort project. When this unintended decision came to light, instead of blaming the new chairperson, who understandably felt embarrassed and ashamed, the district assembly decided to impose unrealistically stringent conditions for the approval. The most notable condition was the demand for the construction of a massive

200-meter-wide mudslide-control dam along the Ashima River to protect the district from potential flooding. The locals referred to the proposed dam as "Nishihara's Great Wall." The assembly chose not to request monetary compensation, as they believed they would have to compromise as soon as the company offered the requested amount. In contrast, the other affected districts, Gōdo and Minami-Gōdo, demanded monetary compensation totaling 40 million yen (about 133,000 U.S. dollars at that time).

Little progress was made in the negotiations for the golf resort project from the fall of 1975 onward. Mayor Takada resigned due to illness (rumors suggest it was partly due to the excessive strain from the conflict). On March 7, 1976, Tomio Ōta was elected as the new mayor. He made aggressive efforts to mediate the negotiations, and by 1977, Sankyō had accepted most of the requests from the affected districts, except those of Nishihara. Eventually, a study sponsored by the village and conducted by an engineering professor at Shinshu University concluded in January 1978 that the dam was unnecessary for flood prevention purposes. With this verdict, all obstacles were cleared, and the project was ready to move forward. However, Nishihara's demand for the "Great Wall" effectively caused a two-year delay in the project.

In the meantime, Sankyō was facing its own difficulties. The oil crisis shook the Japanese economy shortly after the company proposed the golf resort in Matsukawa in December 1972. The company had grown through aggressive real estate development of second homes and, struggling with financial difficulties, was unable to pay the land-holding tax in 1977. On March 20, 1978, two months after the final obstacles to the project were cleared, the Tokyo District Court declared the company bankrupt. As a result, the golf resort project came to a halt. Eventually, the village purchased Sankyō's 61-hectare land for 120 million yen, a significantly lower price than what Sankyō had paid for it. This measure prevented the worst-case scenario of disorderly and fragmented transfer and development of the land.

We highlight three critical roles that the *jichikai* system played in this episode. First, the customary protocol of allowing disproportionate voice to "affected" *jichikai* granted significant political power to Nishihara. In particular, without the *jichikai*, the voice of the "local outsiders" (young migrant settlers) who typically had limited social capital in the village would not have carried as much weight. Their voices gained legitimacy through the *jichikai*, whose membership is based solely on the location of residence. Second, it was the Nishihara district that provided the platform for the knowledge and resources of the MNCS to be effectively mobilized. In other words, the institution of *jichikai* ultimately provided an enduring and politically difficult-to-ignore ground for community resistance that the voluntarily organized civic movement alone could not achieve. Third, *jichikai* provided an organizational framework for the district to address the "unusual" event (i.e., golf resort project) alongside other regular and recurring issues such as repairing damaged walking paths and planning for the next village festival. This also mitigated the risk of alienating the actors of the opposition movement in the village community, which is a critical concern in a small community where regular face-to-face interactions are the norm and nearly inevitable. In these ways, the "old" institution of *jichikai* system critically complemented the activities of the civil society organization MNCS.

Everyday Knowledge of the Village Community

A massive project like the golf resort development in a small rural community has the potential to deepen social divides, erode mutual trust, and give rise to political corruption due to the strains of disagreements, conflicts of interest, and illicit money flows associated with such a project, regardless of whether it is eventually built or not. The *jichikai* system offered a way to mitigate potential social distress by "normalizing" resistance, as described above. However, our fieldwork also led us to believe that there was something beyond formal institutions that somehow "held the community together." The concept of "everyday knowledge" is relevant in this context. Let us present two illustrative episodes to elucidate this further.

First, as mentioned earlier, Tada played a crucial role in spearheading the resident movement and is widely recognized as a key figure in the controversy. Interestingly, during our interviews, one of Tada's friends, who served as our informant, recalled that it was actually Tada's elder brother who initially expressed concerns about the proposed golf resort project. However, he asked his younger brother to organize and lead the movement. The informant suspected that this decision was influenced by the fact that Tada's elder brother had been a prisoner of war in Siberia during World War II and, like many other prisoners in the former Soviet Union, had become a devoted communist. It was believed that if he had become the face of the movement, it might have been perceived as having a distinct "communist color."

Second, we were intrigued by the way village residents portrayed members of the Japan Communist Party during the controversy. Aside from Tada's connection with his brother, there were other Japan Communist Party members who played significant roles in the MNCS and the opposition movement in general. It was evident that the village community as a whole did not embrace the party. During our interviews, we encountered situations where we essentially asked participants in the movement why they were still willing to collaborate with communists. A typical response was something along the lines of, "Yes, he was a communist, but he was a Matsukawa communist; he prioritized the village's interests over the party's interests per se." This usage of the phrase "Matsukawa communist" (which is not a formally recognized political party) caught our attention.

When looking at these episodes, it may be tempting to interpret Tada's brother's decision as an act of disguise and the village residents' use of "Matsukawa communist" as a lack of political ideological consistency. However, we can offer alternative interpretations of these episodes from the perspective of everyday knowledge, as discussed by Matsuda (2013a, b). In this view, their acts can be seen as expressions of their everyday knowledge, which allowed them to engender and sustain the movement for the interest of the village community as a whole, even though they were fully aware of the political ideological divide among themselves.

This type of knowledge is not formally institutionalized or codified; rather, it is a form of meta-knowledge that guides individuals in "knowing what to say and do when" in their everyday lives. At first glance, the role of everyday knowledge may appear irrelevant to the golf resort conflict itself. It is not the type of objective scientific knowledge (such as the potential risks of water pollution and floods, or the economic benefits and costs) or traditional cultural knowledge (related to sacred rocks and cultural legends of the area) that were explicitly employed, particularly by civil society organizations, in the movement. While these forms of knowledge and the associated discourses were undoubtedly at work, we also believe that everyday knowledge, as exemplified in these episodes, played a subtle, yet crucial, role in the controversy. For instance, if an "outsider" (such as an antidevelopment activist) were to lead the movement without much awareness of this locally grounded everyday knowledge, they might encounter unexpected resistance and reluctance even from their local partners. The understanding and integration of this everyday knowledge were essential in navigating the dynamics of the community and facilitating effective collective action.

It is not within the scope of this chapter to determine the specific factors responsible for cultivating such everyday knowledge. However, we suspect that it largely arises from shared life experiences within the community. Traditionally, rice production required close cooperation among farmers, involving the sharing of water, labor, and sometimes equipment across different areas. Nishihara's residents also regularly collaborated with farmers from outside their district. Additionally, Matsukawa is one of the small number of towns and villages in Japan that have never undergone any municipal mergers since the Great Municipal Merger of Meiji in 1889.

This historical context, combined with the absence of clear internal geographic divisions (such as major valleys or hills separating settlements), contributes to the cultivation of a sense of unity and shared destiny among village residents, even across districts that were relatively autonomous villages until the end of the Edo period. The presence of only one elementary school and one junior high school in the village may have further fostered a strong identity among residents as "Matsukawa villagers." The significance of the "old" village boundaries becomes more apparent when comparing Matsukawa to many other Japanese villages that have undergone recent mergers. In such cases, political issues often quickly transform into inter-district conflicts, highlighting the importance of maintaining a sense of unity within Matsukawa.

Conclusions

Undoubtedly, some luck played a role in "preserving" the Gōdohara area of Matsukawa from the golf resort development. During that time, large-scale construction projects were seen as symbols of modern development throughout the country. The golf resort project may have come to fruition if Sankyō had not gone bankrupt. However, despite these circumstances, the case study offers valuable insights into how the local community dealt with the pressures of development.

Both modern civil society organizations and "premodern" local institutions, such as the *jichikai* system, played complementary roles, supporting the argument put forth by Chen (2010) regarding the continuing significance of *minjian* in Asia. The voluntary participation of village residents in the local civil organization, MNCS, had a significant impact during the controversy. However, it was not enough to sustain a prolonged resistance against the development. In retrospect, even the formally democratic institution of the village council did not produce a satisfactory outcome.

In this particular case study, the *jichikai* system, which granted disproportionate political power to specific areas of the locality, emerged as a significant factor. It lent legitimacy to the voices of those who were likely to be most affected by the project but were socially marginalized, such as the new settlers in Nishihara. Furthermore, it helped to sustain the opposition movement by incorporating it as part of the regular business of the Nishihara District.

Finally, our study provides empirical insights into the concept of everyday knowledge, as discussed by Matsuda (2013a, b). Simply put, while various scientific and traditional knowledge was employed by stakeholders during the controversy to advance their respective agendas, we also acknowledge the subtle, yet critical, role played by the local community's everyday knowledge. This knowledge helped mitigate social distress and the negative consequences arising from the project and the opposition movement itself. Viewing the situation through the lens of everyday knowledge allows us to move beyond the publicly expressed discourses and frames the issue not solely as a battle for or against the development project. Instead, it prompts us to consider how the village community maintained a degree of normalcy and cohesion amid the socially divisive external forces.

Notes

1 The original fieldwork for this study was conducted in 2011–2012. At that time, the findings were framed within the literature of community resilience (Yamamoto and Yamamoto 2013), and a life-environmentalist perspective was not foregrounded in the article. This chapter reframes the study in the theoretical context of life-environmentalism with revised interpretations.
2 Motoji Matsuda originally introduced the term *seikatsu-chi* ("life knowledge") in his Japanese text in 1989, and he continues to use it in his Japanese writings (e.g., Matsuda 2013a). However, when writing in English, he opts for the term "everyday knowledge" (Matsuda 2013b). In this chapter, we also use "everyday knowledge," but it should be noted that the emphasis on the expedient aspect of the knowledge is not apparent in Torigoe's (1989) use of the term "everyday knowledge" (*nichijo-chi*).
3 This does not mean that the selected discourse is simply a random or baseless choice. Rather, as it becomes legitimized and essentialized, it gains reality in their everyday life-world and provides stable identities for the residents (Matsuda 2013a, b).

References

Bruchac, Margaret M. 2014. "Indigenous Knowledge and Traditional Knowledge." In *Encyclopedia of Global Archaeology*, edited by Claire Smith, 3814–24. New York: Springer Science and Business Media.
Chen, Kuan-Hsing. 2010. *Asia as Method*. Durham and London: Duke University Press.
Elmhirst, Rebecca. 2015. "Feminist Political Ecology." *The Routledge Handbook of Political Ecology*, edited by Gavin Bridge, Tom Perreault, and James McCarthy, 519–30. London: Routledge.
Furukawa, Akira, and Motoji Matsuda. 2003. *Kankō to Kankyō no Shakaigaku* [Sociology of Tourism and Environment]. Tokyo: Shinyōsha.
Gadgil, Madhav, Fikret Berkes, and Carl Folke. 1993. "Indigenous Knowledge for Biodiversity Conservation." *Ambio* 22 (2/3): 151–56.
Huang, Yi-Chen, Mei-Fang Fan, Chih-Yuan Yang, and Leslie Mabon. 2020. "Social Science Studies of the Environment in Taiwan: What Can the International Community Learn from Work Published within Taiwan?" *Local Environment* 25 (1): 36–42. https://doi.org/10.1080/13549839.2019.1693987
Matsuda, Motoji. 1989. "Histuzen kara Bengi e – Seikatsu Kankyō shugi no Ninshikiron" [From Inevitability to Expediency: Epistemology of Life Environmentalism]. In *Kankyō Mondai no Shakai Riron: Seikatsu Kankyō Shugi no Tachiba Kara* [Social Theory of Environmental Problems: From the Standpoint of Life-environmentalism], edited by Hiroyuki Torigoe, 93–132. Tokyo: Ochanomizu Shobō.
———. 2013a. "Gendai Sekai niokeru Jinruigakuteki Jissen no Konnan to Kanōsei" [The Difficulties and Potentials of Anthropological Practice in the Contemporary World]. *Bunka Jinrui Gaku* 78 (1): 1–25.
———. 2013b. "The Difficulties and Potentials of Anthropological Practice in a Globalized World." *Japanese Review of Cultural Anthropology* 14: 3–30.
Mazzocchi, Fulvio. 2020. "A Deeper Meaning of Sustainability: Insights from Indigenous Knowledge." *The Anthropocene Review* 7 (1): 77–93. https://doi.org/10.1177/2053019619898888
Torigoe, Hiroyuki. 1994. *Chiiki Jichitai no Kenkyu* [Study of Local Self-Governing Associations]. Kyoto: Minerva Publishing.
———. 1997. *Kankyō Shakaigaku No Riron to Jissen: Seikatsu Kankyō Shugi No Tachiba Kara* [Theory and Practice of Environmental Sociology: Perspectives of Life Environmentalism], Tokyo: Yūhikaku.
Wilson, Geoffrey Alan. 2012. *Community Resilience and Environmental Transitions*. New York: Routledge.
Yamamoto, Daisaku, and Yumiko Yamamoto. 2013. "Community Resilience to a Developmental Shock: A Case Study of a Rural Village in Nagano, Japan." *Resilience* 1 (2): 99–115. https://doi.org/10.1080/21693293.2013.797662
Zarsky, Lyuba, and Simon S C Tay. 2000. "Civil Society and the Future of Environmental Governance in Asia." In *Asia's Clean Revolution Industry, Growth and the Environment*, edited by David Angel and Michal Rock, 128–54. London: Routledge.

Part II
Governing Everyday-Life Spaces

6 From Dichotomous Interpretations to Spectrum Thinking

Formation of a Community Organization in a Nuclear Host Locality

Atsushi Yamamuro

Community and Division in Nuclear Host Localities

Residents of localities that host nuclear-related facilities face various difficult challenges. These include ensuring the safety of facilities when they are operating, the formulation of evacuation plans in times of accidents, and coping with regional economic distress when facilities are closed, all of which can significantly affect the course of the lives of local residents. Therefore, if a nuclear host locality has, or can form, a strong community organization of stakeholders to collectively tackle these challenges, it can contribute considerably to sustaining and improving the quality of residents' everyday life. However, in reality, the formation of such a community organization is often impeded or its activities are easily compromised in nuclear host localities.

One reason for the difficulties in forming such organizations is the socially divisive nature of nuclear energy development in Japan (and possibly elsewhere). Typically, we assume that local social divisions are the matter of conflicting opinions, either "for" or "against" nuclear energy or any problems associated with its facilities. However, I argue that critical divisions arise not between those residents who are "for" and those who are "against" the issue, but rather, between those who express their views and those who avoid expressing their views (whether their views are for, against, or otherwise).

This division between residents who express and avoid expressing their views is the result of various policies surrounding the development of nuclear energy and the special features of nuclear host localities. In Japan, local residents are often barred from opportunities to participate in nuclear policy decisions, in which plans are often decided by others and pushed onto the local residents forcefully or through various conciliatory measures. This closed nature of nuclear-related decision-making results in structures and limits how local residents come to express their views on the issues (Hasegawa 2004). Once nuclear facilities are established, host localities become reliant on "nuclear money," including various subsidies and grants from the national government for municipalities hosting nuclear facilities under the "Three Power Source Development Laws" (*Dengen Sanpō*; see Aldrich (2013) for a fuller discussion)

DOI: 10.4324/9781003185031-8

and tax revenue from those facilities. Furthermore, because many of the local residents are either employed full time or conduct contract work in the nuclear industry, they tend to become economically and psychologically dependent on nuclear facilities.

These institutional and economic characteristics tend to shape residents' views as they face nuclear-related issues and then intensify the conflict of opinions among the residents. At the same time, they encourage some residents to act against their will, as they fear negative consequences from revealing their real thoughts, and others to only selectively express their opinions (e.g., to their close families and friends). The deeper the conflict of opinions becomes, the more those who have been vocal about their views cling to their stances, while those who have avoided disclosing their opinions retreat into ever-greater avoidance. As these divisions between those who speak their opinions and those who avoid expressing their opinions grow ever deeper in these ways, the residents who deal with various community-wide issues and the ways they do so become ever more entrenched.

Nevertheless, if the residents consistently respond to community-wide problems in such a divisive and unproductive way, it can result in inadequate nuclear accident prevention measures or incoherent regional socioeconomic policies, ultimately damaging the quality of residents' lives. Under such circumstances, is it possible for residents to overcome the divisions that arise between those who express their opinions and those who avoid doing so? And, what kind of community organizations can be formed through such efforts?

Such questions have rarely been asked in the existing social scientific studies on Japan's nuclear development and host localities. Many of these studies examine political processes over the siting decisions of nuclear facilities (e.g., Lesbirel 1998; Nakazawa 2005; Aldrich 2008), local antinuclear movements (Hasegawa 2004), or socioeconomic effects of nuclear facilities on host localities (Plummer and Yamamoto 2019). While useful, they typically assume or focus on clearly articulated viewpoints of involved actors (whether "for" or "against" nuclear facilities). Few studies have explicitly focused on the challenge of community organization formation amid the situation where many, if not most, residents avoid openly expressing their views and opinions.

In this chapter I first articulate the challenge faced by the residents of nuclear host localities as the problem of "interpreted objects" based on dichotomous categories. I postulate that a successful formation of community organizations depends on how the residents overcome this problem. To empirically illustrate such a process, I focus on a conflict that arose in Tōkai Village, Ibaraki Prefecture, surrounding the installation of a low-level radioactive waste incinerator (hereafter, the "incinerator project") by JCO Co., a nuclear fuel processing company. I examine how the residents have overcome the divides that arose from the incinerator project through the process of the eventual formation of a local community organization, the Air Dose Committee, that keeps a check on various issues related to JCO.

Analytical Framework: How Do Residents Express Their Views?

Life-environmentalism research offers critical baseline insights to explain key dynamics of how the splitting of residents around environmental issues comes to be. To put it succinctly, life-environmentalism points out that in most environmental conflicts residents' individual thoughts and feelings are converged into a few group-based "logics of persuasion," which are called *iibun* in Japanese (Torigoe 1997, 2014). Importantly, those few logics of persuasion are not identical to the residents' original thoughts and feelings (e.g., some of the idiosyncratic feelings may be "stripped off" from the more publicly shared group-based logics as they may not appear "rational" to most other people). Nevertheless, the proponents of life-environmentalism argue that the residents begin to embrace and express their group-based logics as their own. Furthermore, life-environmentalism cautions analysts not to excessively assume resident actors' attributes (age, gender, occupation, political orientation, etc.) as the main determinants of these group-based logics, as many sociological studies tend to do, whether explicitly or implicitly. Instead, life-environmentalism postulates that *iibun* are shaped and controlled by residents' active agency and the knowledge they can use to safeguard their everyday lives, or, as Matsuda (2009) calls it, agency and knowledge used for the purpose of "the convenience of carrying out everyday life."[1]

This formulation of group-based logics of persuasion, while insightful, does not clearly account for the oft-observed reality of residents' avoidance of their views and opinions. Accordingly, I argue that we must first articulate the problem of "interpreted objects," in which residents of a nuclear host locality are constantly and inevitably interpreted by others (Inoue 1977, 204). Understanding the problem of "interpreted objects" is critical because "interpretations" by outsiders can result in intensifying the prevailing division among the residents who try to resist manipulative interpretation by others, but end up reinforcing existing discourses. Moreover, residents' avoidance to express their disagreement over a particular issue may be unilaterally interpreted by outsiders as tacit "consent" (i.e., if they do not explicitly say that they are against a project, they may be interpreted that they are for it).

The analysis of nuclear host localities therefore must include a consideration of the effects of "being interpreted." Here I bring our attention to situations in which residents of a nuclear host locality are interpreted based on dichotomous questions by various others. These outsiders ask their questions in a way that allows for only one of two possible answers. Dichotomous questions that may arise in the case of a nuclear accident, for example, include whether agricultural products are safe or not and whether nuclear energy should be supported or not. Three things happen when residents are faced with such dichotomous questions.

First, the most important thing to note here is the power that these dichotomous questions have to unwillingly and forcefully pull a person into one of the two categorized groups, irrespective of whether that person even actually publicly answers the question. As Seiyama (2000) argues, just the fact of being asked a question in this dichotomous way unwillingly pulls respondents into a newly opened "world of meaning," in which the questions become "meaningful" to – in the sense of having real impact on – their lives (Seiyama 2000, 159). Seiyama calls this process "subsumption through meaning." In the context of nuclear siting, residents of a nuclear host locality are not allowed to give vague answers to outsiders' dichotomous questions (e.g., "Are you for or against nuclear energy?"), and the moment they do answer such a question, a new world of meaning rises up to envelop them. In this new world of meaning, their relationships with other residents and other aspects of their everyday lives could be severely affected. The more sincerely respondents answer outsiders' dichotomous questions, the more powerful the pull into this new world of meaning becomes.

Second, as residents' responses are churned through this dichotomous categorization, the residents themselves become "interpreted objects" (Inoue 1977, 204). Becoming an "interpreted object" effectively means that someone has the intentions of their words and actions twisted by others (they are effectively *mis*interpreted), but then they nevertheless are eventually forced to act according to these (mis)interpreted meanings. The asker of the dichotomous question will unilaterally continue the process that leads to misinterpretation even if the "object" being interpreted (the respondent) no longer responds and remains silent.

Third, residents' responses to one dichotomous question may call forth other dichotomous questions. For example, what starts as a conversation about the safety of agricultural products (e.g., after a nuclear accident in their locality) can morph into a question of whether the respondent approves or disapproves of nuclear energy.

To recap, the three key processes pertaining to dichotomous questions that arise in nuclear host communities are as follows:

1 Due to the very nature of such questions, respondents are subsumed into a new "world of meaning" unleashed by the questions.
2 Once subsumed into this world of meaning, there is a possibility that they will be made into "interpreted objects."
3 There often arises a kind of chain reaction of dichotomous questions.

As these three processes mutually reinforce each other, they intensify the divisions between residents over whether or not to express their thoughts and feelings about the focal issue.

So, what could residents possibly do to overcome these divisions? The key is to create a new category that relativizes the existing dichotomy. In order for this to happen, residents need to speak from their true "spectrum" of views, relentlessly reviewing and scrutinizing the inadequacy of the dichotomous categories through which they are interpreted, and repeatedly formulate new and reformulate old categories (Takenaka 2008, 39). That is, residents will need to recast the issue at hand, which has been made into a dichotomized "either-or" question, as an issue that embodies a range of views and positions that are connected to each other in a kind of spectrum. I call this kind of thinking, which allows residents to refuse being dichotomously categorized, *spectrum thinking*. Spectrum thinking creates new categories to better articulate their group-based logics that resonate with their lived experiences. In this way, residents can seek to overcome internal divisions among themselves by creating mechanisms to express their thoughts and intentions through spectrum thinking.

The above analytical framework elucidates how the views of residents living in a nuclear host community are expressed, interpreted, misinterpreted, and can be better understood. I use this analytical framework to examine the case study below.

Local Residents and the Development of the Incinerator Project

Divisions over the Incinerator Project

Tōkai Village (*Tōkai-mura*) is located 15 kilometers northeast of Mito City, the prefectural capital of Ibaraki Prefecture, and has a population of approximately 38,000 (Figure 6.1). In Tōkai Village, the birthplace of Japan's nuclear energy development, there are more than ten nuclear-related facilities, including a nuclear power plant and various engineering firms and research institutes. In 1999, a criticality accident occurred at JCO (Hasegawa and Takubo 2001; IAEA 2009). The accident was caused by careless work by employees who failed to observe state-mandated nuclear fuel processing procedures. As a result of the employees' mishandling of nuclear fuel, nuclear fission chain reactions continued in an unprotected area of the facility for 20 hours. Two employees

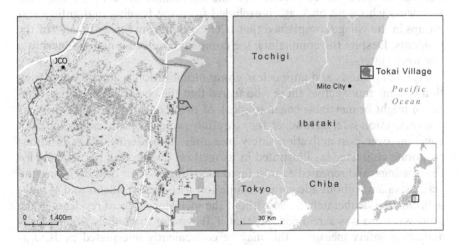

Figure 6.1 Location of Tōkai Village and JCO. Maps by the author.

died from acute radiation effects and 666 people, including residents near JCO, who were suspected of having been exposed to radiation underwent exposure dose assessment. Residents within a radius of 350 meters from the accident site were officially ordered to evacuate and about 310,000 people within a radius of 10 kilometers were ordered to take shelter indoors. This was the first nuclear accident in Japan where residents had to evacuate. In 2003, JCO decided to leave the nuclear fuel processing business and since then, its main business has been the storage and management of radioactive waste.

In February 2012, JCO officially shared its plans for the incinerator project with the Tōkai Village government and began preparations to request a permit from the national government. In June, the company announced the plan to the residents. However, the method of public announcement was seen as problematic, since not all residents were equally given the chance to publicly express their views on the project. Even though this was a brand-new project, which had little to do with the 1999 criticality accident, the company invited only those residents who lived in the 350-meter-radius evacuation zone[2] that was set at the time of the accident in 1999 to the formal informational briefing about the new project. In other words, the company made a unilateral decision to use the binary category of either inside or outside the 350-meter radius of the 1999 JCO Criticality Accident to determine its audience. Residents invited to the meeting expressed concerns and argued against the project. Residents outside the 350-meter radius, who learned about the project through various channels, expressed distrust of JCO and opposition to the construction of the incinerator.

Nevertheless, JCO did not withdraw the plan and went ahead to apply for a permit from the national government. In August, JCO held another formal briefing for residents within the 350-meter radius from the 1999 accident to let the residents know that they had gone ahead with the permit request. Invited residents criticized the ex post facto reporting of the permit application submission, and demanded that the incinerator project be withdrawn. Residents who were once again excluded from the briefing and antinuclear groups in the village complained that JCO was ignoring the sentiments of the residents. Despite the complaints, the company received a national permit a few weeks later.

As local protests and antinuclear group mobilization against the construction became more active, those who feared that the construction of the incinerator might be inevitable began to speak of their concerns about the facility's post-activation safety. However, they carefully avoided openly expressing their concerns over post-activation safety measures. Considering JCO's insidious corporate character (as illustrated in its method of selecting participants for the briefings, and repeated ex post facto reporting of their own decisions and actions, taken without adequate consultation with local residents and at times willfully and deliberately made without the inclusion of local residents in the process), there was a real concern that if they were to publicly discuss post-activation safety measures, this may be conveniently interpreted by JCO as residents' approval of the project. These residents also did not want to give any impression to those fellow residents who suffered from the criticality accident that they approved of the new project.

And so, even as the formalities necessary for moving ahead with the construction were being pushed through by JCO in their negotiations and paperwork with the national government, on the local level, there came to be a growing division between those who openly opposed the construction of the nuclear facility flat-out and those who, even though they wanted to talk about the facility's post-activation operational safety, avoided doing so openly due to the concerns described above.

Emergence and Change of a Local Resident Group

In September 2012, a group calling itself "Local Residents of the Surrounding Area" presented JCO with a formal written appeal. The group was led by A-san, who was 60 years old at the time and lived 300 meters from the JCO facility. They described themselves as "Local Residents of the Surrounding Area (*shuhen jūmin*)." The intention behind this general self-categorization was to offer a counter-narrative to JCO's unilateral decision to use "inside or outside the 350-meter-radius from the site of the 1999 criticality accident" as the means to define relevant stakeholders. These "Local Residents of the Surrounding Area" argued, instead, that "there is no [such] line [demarcating 'inside' and 'outside'] drawn on the ground." This understanding was widely shared among the local community and informed by their experience during the criticality accident, when an elementary school 800 meters away from the scene was notified of the accident only after a considerable delay, which in turn caused the school's response to protect the children to also be delayed. That is why A-san and the members of the group took it upon themselves to visit homes in the surrounding community, including those outside of the 350-meter radius, to earnestly listen to residents' candid concerns when creating the written appeal that they presented to JCO.

The appeal demanded an overhaul of JCO's corporate culture and enumerated a series of safety measures that they expected to be put in place if the incinerator was to ever be constructed. It went on to state that if JCO failed to meet these demands, then "we, as selected residents who live within 350 meters, will oppose the proposed project to the bitter end." As residents who were formally notified by JCO about the project, A-san and some others feared that unless they openly expressed their distrust of JCO and their concerns about the project, JCO might one-sidedly interpret that the residents had "accepted" the project. In order to avoid becoming an "interpreted object" and to bring safety issues to the table, they intentionally used the categories of "opposition (to the project)" and "within 350m (from JCO)."

From October to November 2012, at the request of Tōkai Village, JCO held three briefing sessions targeting the four *jichikai* (neighborhood associations) adjacent to the firm.[3] The holding of these sessions effectively expanded the areal definition of targets for explanation beyond the 350-meter radius, and created the atmosphere that neighborhood associations near JCO were legitimate agents of negotiation. At one of the briefing sessions, a new fact was discovered. In response to a question from a participant, it came to light that radioactive waste generated by the criticality accident in 1999 was still being stored on the premises of JCO. It was at that moment that members of A-san's "Local Residents of the Surrounding Area" group who participated in the briefing became painfully aware of JCO's insidious corporate culture, steeped in cover-ups and lies. They became deeply worried that if they simply expressed opposition to the project, it would not be possible to cope with various problems that might arise in the future, if the project was forced through anyway. However, the briefings played like a broken record: as antinuclear groups proclaimed from start to finish of each meeting that they were against the construction, JCO would counter with their rehearsed explanation that the incinerator was safe and necessary.

One week later, A-san's group changed its name from "Local Residents of the Surrounding Area" to "Residents Demanding Safety (*anzen*) of Operations and Peace of Mind (*anshin*)." They also changed their position from being "against" the construction of the facility to "we do not necessarily seek to have the project frozen or abandoned" and submitted a list of six demands to JCO. One of the demands asked that residents be included and present when inspections of the incinerator take place. When asked why they changed their group name and positions, A-san stated:

> For us to have peace of mind, we as local residents have to confirm for ourselves that the facility is being operated safely. Furthermore, our scrutiny of JCO will also serve as the best safety measure for workers at the facility. There's no need for JCO and us to become enemies.
> (Personal communication, September 20, 2015)

In revising their position, the "Local Residents" of A-san's group drew on their experience of the criticality accident, which was caused by JCO's illegal operating practices. They especially remembered the deaths of the workers and the tremendous damage that the accident inflicted on the local residents. It is from this standpoint that they revised the dichotomy that residents are supposed to be either "for" or "against" the construction and created the new category of "safety and peace of mind for residents *and workers*" [emphasis added]. The direct participation of residents whenever the incinerator would be inspected was one of their demands based on this new category. The group thought that such direct participation in the inspections would enhance the sense of safety among the residents and help ensure safe operation of the incinerator, which would also relieve safety concerns among employees. In the meantime, before formally responding to the group's demand, JCO proposed to "consider setting up a third-party organization to ensure the peace of mind for the residents before the construction of the facility" (as a response to another civil society organization that argued for the importance of such an organization). For

A-san's group, which demands the direct involvement of residents in the inspection, this was an unsatisfactory proposition.

Establishment of the Air Dose Committee

From that point onward, JCO and the Tōkai Village government began to include local neighborhood associations (*jichikai*) in negotiations surrounding plans for the incinerator facility. In February 2013, the Nuclear Energy Section of the Tōkai Village office hosted a "Residents' Nuclear Energy Roundtable" focusing on the incinerator project.[4] The participants included JCO and board members of one of the *jichikai* adjacent to JCO (the First Division of the Funaishikawa area). A-san, one of the members of the Funaishikawa neighborhood association, also participated. At the meeting, it became apparent that the Tōkai Village office "felt that the understanding of the local residents was not progressing" regarding the incinerator project, and that the office was requesting JCO to consider "disclosing information through a third party organization" among other things.[5] These initiatives by the local government and JCO effectively promoted the presence of *jichikai* as legitimate negotiating agents. This did not mean, however, the diminished influence of the "Residents Demanding Safety of Operations and Peace of Mind" group. The First Division of Funaishikawa neighborhood association gathered information about the group's activities from A-san, and eventually appointed him as the main negotiator with JCO.

In July 2013, JCO formally responded to the list of six demands by the "Residents Demanding Safety of Operations and Peace of Mind" group. In response to the demand that residents be included and present when inspections of the incinerator take place, JCO proposed to set up a commission to "confirm" the incineration process to which it would invite experts and board members of *jichikai*. However, the residents' group could not approve such a response. First of all, it did not allow for the direct inclusion of residents in the actual inspection of operations; it only proposed a commission to "confirm" operations at the incinerator. Secondly, participants in this meeting would be limited to *jichikai*'s board members. Accordingly, A-san negotiated directly with JCO and the Nuclear Safety Section of the village office over the course of several meetings. As a result, they agreed to establish the "Air Dose Committee," which would meet on a regular basis and directly include local residents in the measurement of air radiation dose rates on the JCO premises. The construction of the incineration facility then started in January 2014.

The Air Dose Committee, which started its regular meetings in December 2014, conducts surveys four times a year. Participants include representatives from JCO and the village office, up to two people from each of the four neighborhood associations of the areas adjacent to JCO (Figure 6.2) with at least one participant per neighborhood being a member of the "Residents Demanding Safety of Operations and Peace of Mind" group, and up to eight additional participants, regardless of their place of residence. Dose measurements are

Figure 6.2 Four neighborhood associations of the areas adjacent to JCO. Map by the author.

taken at four locations within the JCO premises and the results are published on the JCO website and circulated in the "News from JCO" section of the four neighborhood associations' information boards. The Committee acts as an important communication channel for residents, helping to reduce their anxiety about the operation of the incinerator and other issues pertaining to JCO.

Local Residents' Spectrum Thinking

Let us review how the Air Dose Committee came to be, keeping in mind the analytical discussion at the beginning of this chapter. To begin, "Local Residents of the Surrounding Area" revised a series of dichotomies that were imposed on them when it came to the incineration plan. First, they rejected the categories that JCO made for them, which tried to split them along the lines of "inside / outside a radius of 350 m from JCO." They then saw that it was also important to avoid the "for" or "against" dichotomy that was being used by antinuclear groups and JCO. Instead, they came up with two categories of their own.

The first was a category of "Local Residents of the Surrounding Area" that made sense in the local context and that was not beholden to JCO's unilateral decision to use the 350-meter radius. Based on their past experience of the criticality accident, "local residents" are not divided by some sort of "line [demarcating 'inside' and 'outside' the 350 m radius around the 1999 criticality accident] drawn on the ground."

The second was a category of "Safety of Operations and Peace of Mind for the Residents and Workers." Even though the construction of the incinerator was about to go forward, the residents could not negotiate even the safety measures because they were (against their will) interpreted and divided along the lines of "for" or "against" the plan. By returning to the collective experience of suffering by the workers and residents during the criticality accident, they arrived at the realization that there was no clear line dividing workers and residents, and that "there is no need for JCO and residents to become enemies."

These categories devised based on the spectrum thinking opened up a new world of meaning that was different from the old world of meaning based on the existing dichotomy surrounding the incinerator project. As the residents gradually embraced the new world of meaning, they were able to overcome the divide among themselves.

That said, there was still a possibility that the actions of A-san's group, with its spectrum thinking, could have been criticized and rejected by residents who lived close to JCO. The way that JCO was going about its plans for the incinerator could well have reminded them of their suffering at the time of the criticality accident and could easily have evoked a victim–perpetrator dichotomy. Nevertheless, the group gained support from locals who lived near JCO, as well. How come? For an answer, we need to look closely at the Mr. and Mrs. A's active local engagement that began at the time of the criticality accident and continued thereafter.

A-san and his wife run a food processing business 300 meters from JCO. After living through the criticality accident, they had no choice but to give up on their plans of passing on the family business to their son. They gave up on this plan because after the accident, they realized how difficult it was to run a food-related business somewhere that also hosts nuclear facilities. For example, their customers from all across Japan unilaterally canceled their orders. However, their son unexpectedly told them, "I want to keep living in Tōkai Village," and they could not easily reject him with something like, "but it's dangerous here." Furthermore, they regretted that despite living so close to JCO, they had not at all been interested in anything having to do with JCO until living through the criticality accident first-hand. A-san used this experience as a jumping off point and committed himself to working together with other local residents to think about what kind of Tōkai Village they wanted to hand over to the next generation.

One year after the criticality accident, A-san and his wife began a local initiative that they called "The Ring of Life." They raise flower seedlings, give them out to schools and residents in the area for free, and care for them as they grow. They pay for all expenses out of pocket. They do this because they feel they have fallen into an economic and psychological dependence on nuclear facilities and they want to show their desire to escape from this reliance. They decided that they would focus their efforts on areas within "about two kilometers of JCO," a distance that they can easily and frequently cover whenever they have a free moment in their day-to-day duties with the family business. This area includes the four neighborhood associations who came to be included in the Air Dose Committee. Whenever they talk with people as part of this initiative, A-san and his wife try their best to get people to think and imagine broadly what the future of Tōkai Village might look like and try not to box people in based on their stance in favor or against nuclear power. They do so because they know that there will be people who avoid expressing their views if presented with this dichotomy. They also started volunteering together cleaning sidewalks with a volunteer group they met. All of these efforts have garnered A-san and his wife a local reputation as "people who contribute to the good of our community." People in the community adore them and often come to visit them at their home to discuss the future and present of Tōkai Village. A-san and his wife's efforts, in which they interact deeply and without discrimination with all residents, embody spectrum thinking. Many residents sympathize and work together with A-san and his wife, and their supporters, forming and expanding a strong network.

Even though they began the spectrum-thinking-based activities of the "Local Residents" group at a time when residents were becoming divided on the incinerator project, because A-san and his wife had developed trust-based relationships with many residents, the group was not seen to be employing sophistry or siding with JCO. To the contrary, people widely supported A-san's group precisely because they hoped that it would have the power to make "the inclusion of local residents in the inspection of incinerator operations" a reality, a vision that was a concrete manifestation of spectrum thinking. In this way, residents were able to overcome the local divisions surrounding the incinerator project and at the same time gain an organizational channel to express their views in the form of the Air Dose Committee.

Conclusions

The residents of Tōkai Village had felt a need for some sort of community-based organization to deal with problems arising from JCO. With the establishment of the Air Dose Committee, a definite shape was finally given to such an organization. They now understand that the proper constituencies for interacting with JCO on these matters are the residents of the four neighborhood associations and the areas they encompass. It has at its base the four neighborhood associations surrounding JCO and provides a mechanism for these constituencies to express their views based on spectrum thinking.

I point to two key factors that played critical roles in the formation of the Air Dose Committee. The first is that the newly formed community organization, enabling the expression of residents' views, still functions in conjunction with traditional neighborhood organizations (*jichikai*), highlighting the continuing importance of preexisting local institutions. The second is that A-san and his wife's activities that were based on spectrum thinking, deeply informed by their lived experiences, leading to the accumulation of social capital on their behalf, were essential in the successful establishment of the Air Dose Committee.

Because of the formation of the community organization, it became difficult for JCO to make unilateral decisions when it came to their construction plans or who would be the target audience at their informational briefings, like they used to. In other words, this organization has made itself a presence that JCO must acknowledge and incorporate into its decision-making process. It is a community organization that has gained a legitimated voice and certain discretionary rights based on regular monitoring of JCO's activities. Accordingly, this community organization can help sustain a continuous engagement of the residents with various problems arising from nuclear facilities with some degree of effective power, thereby securing everyday life in nuclear host localities for years to come. It goes without saying, nevertheless, that to maintain the functioning organization, it is essential that the residents continuously revisit and update as necessary the key mechanism to express their views drawing on spectrum thinking. In the context of this study, the residents may need to modify the Air Dose Committee when they face a new JCO-related problem. This, in turn, requires local residents to continue everyday life activities that are rooted in spectrum thinking.

Notes

1 Here "convenience" does not mean that the residents will choose their actions because something makes their life materially easier and more comfortable. Rather, it means that they handle situations in such a way that will be most effective and at the same time cause the least amount of trouble and tension, taking a myriad of factors that may affect their everyday life into consideration.
2 The actual evacuation zone encompassed blocks of land that loosely traced a circle with a radius of 350 meters around the accident site.
3 Editors' note: Torigoe (1994) offers extensive discussion of the characteristics of *jichikai* organizations in Japan. See also Chapter 5 for brief explanations.
4 The Residents' Nuclear Energy Roundtable started in 2008 for the purpose of exchanging opinions with Tōkai Village residents, nuclear power establishments, and the local government, and has been held twice a year with the rotating participation of *jichikai* in Tōkai Village.
5 Based on the reports of the Residents' Nuclear Energy Roundtable, compiled by the Nuclear Safety Section of the Tōkai Village office.

References

Aldrich, Daniel P. 2008. *Site Fights: Divisive Facilities and Civil Society in Japan and the West*. Ithaca, NY: Cornell University Press.
———. 2013. "Revisiting the Limits of Flexible and Adaptive InstitutionsThe Japanese Government's Role in Nuclear Power Plant Siting over the Post-War Period." In *Critical Issues in Contemporary Japan*, edited by Jeff Kingston, 78–91. London: Routledge.
Hasegawa, Koichi, and Yuko Takubo. 2001. *JCO Criticality Accident and Local Residents: Damages, Symptoms and Changing Attitudes: Data and Analysis of the Results of*

a *Field Survey of Tōkai-Mura and Naka-Machi Residents*. Tokyo: Citizens' Nuclear Information Center. https://cnic.jp/english/publications/pdffiles/jco_residents_font.pdf

Hasegawa, Koichi. 2004. *Constructing Civil Society in Japan: Voices of Environmental Movements*. Tokyo: Trans Pacific Press.

IAEA. 2009. "Lessons Learned From the JCO Nuclear Criticality Accident in Japan in 1999." http://www-ns.iaea.org/downloads/iec/tokaimurareport.pdf

Inoue, Shun. 1977. *Asobi no Shakaigaku* [Sociology of Play]. Kyoto: Sekai Shisō Sha.

Lesbirel, Hayden S. 1998. *NIMBY Politics In Japan: Energy Siting and The Management of Environmental Conflict*. Ithaca, NY: Cornell University Press.

Matsuda, Motoji. 2009. *Nichijo Jinruigaku Sengen! Seikatsu Sekai no Shinso elkara*. [Declaration of Everyday Anthropology! To/From the Depth of the Life World]. Kyoto: Sekai Shisō Sha.

Nakazawa, Hideo. 2005. *Jūmin Tohyō Undō to Rōkaru Rejīmu – Niigata Ken Maki Machi to Kongen Teki Minshushugi no Hosomichi, 1994–2004* [Resident Voting Movement and Local Regime: Maki Town, Niigata Prefecture, and the Narrow Path of Radical Democracy, 1994–2004]. Tokyo: Harvest.

Plummer, P., and Daisaku Yamamoto. 2019. "Economic Resilience of Japanese Nuclear Host Communities: A Quasi-Experimental Modeling Approach." *Environment and Planning A* 51 (7): 1586–608. https://doi.org/10.1177/0308518X19852125

Seiyama, Kazuo. 2000. *Kenryoku* [Power]. Tokyo: University of Tokyo Press.

Takenaka, Hitoshi. 2008. *Jiheishō no Shakaigaku – Mōhitotsu no Komyunikēshon Riron* [Sociology of Autism – Another Communication Theory]. Kyoto: Sekai Shisō Sha.

Torigoe, Hiroyuki. 1994. *Chiiki Jichikai no Kenkyu* [Study of Local Self-Governing Associations]. Kyoto: Minerva.

———. 1997. *Kankyō Shakaigaku No Riron to Jissen: Seikatsu Kankyō Shugi No Tachiba Kara* [Theory and Practice of Environmental Sociology: Perspectives of Life Environmentalism]. Tokyo: Yūhikaku.

———. 2014. "Life Environmentalism: A Model Developed under Environmental Degradation." *International Journal of Japanese Sociology* 23 (1): 21–31.

7 "Public" (*gong*) as Village Norm
Urbanization and Community Response in China

Meifang Yan

Rural Urbanization Policy in China

Farming villages of northern China have been considered by researchers as having a weak sense of community (Fukutake 1976, 490). This implies that villagers share few common ethical principles or norms, collectively called *gong* (公) in Chinese, that are specific to their villages. This chapter counters this conventional understanding by describing a case where a "community *gong*" arose when the village faced a crisis and the villagers' lives were at risk. More broadly, this chapter provides an empirical case study of life-environmentalism in a non-Japanese context.

The economic disparities between cities and rural villages are growing in China, and the government is rolling out the rural urbanization policy (农村城镇化政策: "farming village urbanization policy") nationwide in its effort to reduce the disparities. The policy involves three "concentration" initiatives: concentration of population into modern residential zones, of industry into economic development zones, and of farmland management rights into farmers' hands. Specifically, integrated development of agriculture and industry is being sought by, on one hand, encouraging farmers to relocate from villages to modern-style housing complexes, and on the other hand, by converting sites no longer used for rural housing into industrial land. Farmers then commute from their housing complexes to work in the reorganized agricultural land; they also have the option of earning non-agricultural income in newly built factories and other businesses. As a result, the scenario goes, an increasing number of farmers will relinquish their land, leading to the consolidation of agricultural land and the development of large-scale farms (Yan 2010).

The rural urbanization policy has led to pressure on villagers to abandon their homes, causing the disappearance of more and more villages across China. For example, in the Wuqing District of Tianjin City, seven villages were dissolved after their residents were all relocated into one housing complex (Yan 2019).

However, there are villages, such as Village X examined in this chapter, that have resisted the state policy by using the village-level, "community *gong*" as a justification. How have the residents of Village X been able to work together and sustain their resistance? In order to answer this question, this chapter focuses on how the relations among the villagers in the time of crisis evolved through their daily interactions, giving rise to the "community *gong*." This study will in turn shed light on the formation of the "community *gong*" in contemporary China.[1]

Conceptualization of *gong* (Public) and *si* (Private) in China

The Chinese words 公 (*gong*) and 私 (*si*) are typically translated as "public" and "private" in English. However, *gong* and "public" are not synonymous, nor are *si* and "private." Yuzo Mizoguchi, a Japanese historian specializing in Chinese philosophy and culture, conceives of *gong* as having three meanings: political *gong*, *gong* as connections, and ethical *gong* (Mizoguchi 1996, 2001). The first is *gong* as a political quality possessed by leaders such as nobles and governmental authorities (as in "public office"). The second is a communal quality of collectiveness and openness (as in "public space"). Mizoguchi articulates the subtle nuance of the second *gong*, saying that this *gong* refers to the bonds among horizontally connected individuals (*si*). In other words, *gong* applies to concrete private individuals (*si*), rather than the collective "people" or "all" (as implied by the English word "public"). In this conception of *gong*, the "private" (*si*) maintains the right to use or possess the "public" (*gong*) entity. That is to say, in the Chinese *gong-si* perspective, it is a matter of course that public property is used for private purposes.

The third, ethical *gong*, focuses on universal ethical principles with the connotation of equity and impartiality, which, Mizoguchi (1996, 57) suggests, is rooted in the Chinese principle of *tianxia* (天下: "Under heaven"). Chinese anthropologist and sociologist Fei Xiaotong ([1948] 1992) argues in a similar vein that there is a social norm in China, called *li* (礼), which overlies the self, and is similar to the third *gong*. *Li* is manifested as rules and manners that are embedded in various everyday situations and interpersonal relationships. Fei considers *li* as a kind of general law in the society ([1948] 1992, 95).

Beyond these three meanings of *gong*, life-environmentalism, developed originally in the Japanese context, points to another possible type of *gong* (kō in Japanese), which may be called "community *gong*" or "village-level *gong*." In Japanese local communities, especially rural areas, there is a sense of dual-layered land ownership, where "my land" is layered over "our land." The "community *gong*" here refers to the norms of the local community that inform and limit the actions of individual community members. Over time, Japanese local communities developed distinct community *gong* that is distinct from state-level "public" norms and rules, and it has played important functions in securing and enriching people's lives. For example, in the Toga River basin of Kōbe City, Hyōgo Prefecture, the Toga River Protection Association was formed by several small communities to conserve the riverine environment. Notably, through its activities such as the creation of voluntary conservation rules and regular river cleanups, the association has come to hold a strong "joint right of possession" (*kyōdō senyū ken*) over the river, which is officially under the jurisdiction of the government, to the point where it was able to dam the river to create a children's pool (Torigoe 1997). Moreover, when the local government intends to take any action to the Toga River, it is now expected to gain approval from this community association, indicating that the latter has acquired a "public" character (Torigoe 1997, 71–72). Life-environmentalism research has shown that these types of community organizations, and the norms and rules that they rely on, can play an active role in preserving and enhancing people's living environments.

Do these forms of "community *gong*" exist in China? More specifically, in a time of crisis when the state's village urbanization policy is causing many Chinese village communities to disappear, is there any prospect for the formation of a "community *gong*" that will actively resist the state policy and preserve communities?

Previous studies suggest that community-level cooperation and unity are rather weak in Chinese farming villages; instead, they assert, the "private" realm (*si*) has become more pronounced. For example, in his work on the wartime survey materials, the Japanese legal scholar Michitaka Kaino (1942) finds that there was no proper name referring to all of the land in a village, and that each land parcel is called after the individual owner's name. From these, Kaino suggests that a sense of clear, permanent village boundaries is absent in Chinese villages. In addition, Kaino observes that transactions in land, as in most other commodities, were governed by general economic principles, and that there was no equivalent to Japan's *takamochi honbyakushō*, a class of farming households with economic power and high social status who collect land usage rents from peasant farmers while bearing certain responsibilities. From these observations, he concluded that there was little collective sense of a village community in China, and that the village should be thought as a collection of individuals where meritocracy prevails (Kaino 1942).

Researchers who witnessed the economic growth of China after WWII, and especially after the opening of the country in 1978, came to focus on the expansion of the realm of "private" (*si*) and individuals (个: *ge*) in villages as a key feature of rural social change dynamics. Their studies often referred to the notion of *chaxugeju* (差序格局: "differential-order patterns"), coined by Fei Xiaotong ([1948] 1992). According to Fei:

> In Chinese society, the most important relationship – kinship – is similar to the concentric circles formed when a stone is thrown into a lake…. Despite the vastness [of kinship networks], though, each network is like a spider's web in the sense that it centers on oneself…. Each web has a self as its center, and every web has a different center.
>
> (63)

In this way Fei conceived of Chinese society as a convergence of interpersonal networks or social circles centered on the self, which can be "expanded or contracted according to a change in the power of the center" (64).

This elastic quality of social circles makes the boundary between the "public" (the second *gong*) sphere and the "private" (*si*) sphere a relative one. Put another way, the "public" has a nested structure where a family is *gong* for the private self, and a clan is *gong* for the family. Thus, according to the idea of *chaxugeju*, a person may sacrifice his family for his own benefit, and his family may sacrifice its clan for the sake of the family. For Fei, therefore, the private (*si*) is always greater than the public (*gong*) in Chinese traditional thought. Indeed, Fei asserts that the idea of individualism does not exist, only egocentrism does (67).[2]

For example, in the farming village in northeast China studied by Japanese sociologist Toshikazu Shutō (2003), the village secretary completed road surfacing work by mobilizing resources that he obtained through his personal networks that extended outside the village. However, the village secretary had no particular interest in the public good and welfare of the villagers; rather, the road construction was only to secure his legitimacy as a village secretary. Drawing on this case study, Shutō suggests that "public" works tend to be cosmetic/transitory with arbitrary goals, and that serving the "public" interest (in the sense of the English word) has not taken root as an austere normative value in Chinese villages. Rather, events and activities in a village are often decisively influenced by the unilateral decisions and individual disposition of powerful figures such as the secretary in the case study (Shutō 2003, 176). Shutō's study exemplifies the research on the transformation of societies in post-reform China, which emphasizes how "personified" *gong* (or *si*-centered *gong*), as a distinct characteristic of Chinese social life, has contributed to the social dynamics.

In summary, in their analyses of the concept of *gong* in China, Mizoguchi and Fei articulate three key meanings: *gong* as authority and government, *gong* that assumes private rights and interest in the center, and *gong* as more universal manners and ethics. However, beyond these meanings of *gong*, there has been little discussion of "community *gong*" that guides the behaviors of the self and others within a village-like community, and whose significance has been a focus of life-environmentalism. Accordingly, this chapter investigates Village X in Xintai City, Shandong Province, where the villagers came under pressure from the village urbanization policy to relocate. In particular, I focus on how the villagers attempted to reorganize and reinvigorate their daily village life by rebuilding a local temple (关帝庙: *Guandimiao*), and examine how "community *gong*" and communality among the villagers were generated in that process.

Rural Urbanization as an Emergent Crisis

Overview of Village X and Beginning of Rural Urbanization

Village X is located in greater Xintai City, a coal mining region in central Shandong Province. When surveyed in February 2011, the village had 280 households with the population of 826 people (interview with the village committee).[3]

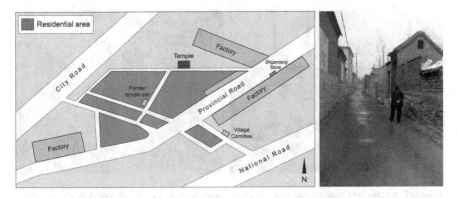

Figure 7.1 Schematic map of Village X and a street in the village. Map and photograph by the author.

Approximately 30% of the villagers engage in agriculture, mainly crop farming, while also working second jobs in coal mining.

Since 1998 Village X began to feel the direct effect of urbanization as the Shandong provincial government started to linearize roads in the province in order to suppress road construction costs. The provincial highway first cut through the village, and the city government subsequently built a trunk road on the western side of the village in 2003 (Figure 7.1). As a result, the village was divided and sandwiched by major roads. Building on these road projects, Village X was eventually included in the urban development zone in the Xintai City Comprehensive Urban Plan (2004–2020). The village was now confronted by village relocation.

Village X was designated for relocation by 2020, with the next step of this plan, following road construction, involving the expropriation of 12.7 hectares (31.4 acres) of agricultural land. Because of this designation, from 2008 onward it was not possible to obtain new land for building houses within the boundaries of Village X. In addition, the village elementary school had already been closed, with a new school covering four villages built approximately 3 kilometers (1.9 miles) from Village X. Construction of housing complexes near the new elementary school to take in villagers from the surrounding villages had already begun.

Construction of a Temple as a Response to Traffic Accidents

After the road construction projects and the inclusion of the village in the Urban Plan, a large number of traffic accidents began occurring in Village X. The provincial highway passing through the village is approximately 20 meters (66 feet) wide and, despite carrying high volumes of traffic, there is not a single set of traffic lights or pedestrian crossing on the road within the village. Since the provincial highway opened, a total of eight people died in traffic accidents in Village X, and after each fatal accident, the villagers petitioned the

Figure 7.2 Rebuilt *Guandimiao* temple, (left), entrance of the temple (middle), and inside of the temple (right). Photographs by the author.

relevant authorities for traffic lights to be installed. However, permission for the provision of traffic lights within Village X was never granted, presumably because of the forthcoming relocation of the village.

After so many fatal accidents, the residents of Village X came to think of the multiple roads cutting through their village as a disruptive presence driving a wedge into the very order of their lives. Complaining that the roads act like "a giant meat cleaver severing their living space in two pieces," villagers began to demand action from the village office. Interestingly, the demand took the specific form of the rebuilding of the temple in addition to such measures as installing traffic lights.

According to a village legend, the *Guandimiao* of Village X is a temple dedicated to Guan Yu, a military general during the Three Kingdoms period (220–280 AD). Due to his devotion to protecting his king with total disregard for his own life, Guan Yu is also a deity widely worshiped in Chinese folk religion for the virtues of loyalty and righteousness as well as safety and success in business. The temple in Village X had a history of over 200 years before being demolished during the Cultural Revolution (1966–1976). Despite the loss of the temple, the residents of Village X continued to regard Guan Yu as a deity that ensures their safety (Figure 7.2).

Still, even if the villagers wished to have the temple rebuilt, budgetary constraints and political considerations could not be ignored. How did the villagers overcome such significant hurdles?

Long Path to Rebuilding the *Guandimiao* Temple

Political Environment Enabling the Reconstruction of the Temple

During the Cultural Revolution, temples (*miao*) were regarded as symbols of "old culture," and were targeted for destruction. After the opening of the country in 1978, the rebuilding of temples was permitted for tourism development, but judgment was suspended regarding the reconstruction of temples specifically for religious worship. Accordingly, the rebuilding of the temple in Village X, which did have an emphasis on worship, could involve a political risk, especially if led by the local administrative office. Hence the initiatives to rebuild the temple were led by a traditional village-level (*minjian*) organization for coordinating funerals, called the *Bailishihui* (白理事会: "White Council").

Before the early 1980s, the village mayor had taken charge of funeral ceremonies. The chief mourner, representing the family of the dead, would customarily hand over all responsibilities, except for the payment of funeral expenses, to the village mayor. In rare cases when the chief mourner was unable to scrape together the funeral costs, the mayor was even permitted to expropriate the family's food or land. This situation began to change in the mid-1980s when the central government directed to discontinue the practice of burial and promote cremations, and to establish a local organization to oversee funeral ceremonies; the *Bailishihui* was then established in Village X. The village mayor of the time, Mr. A, was nominated to be the chair of the *Bailishihui*.

The *Bailishihui* is comprised of three roles: one *Zhishi* (执事) who supervises the work, *Juzhong* (举重) members who supply physical labor (e.g., to build tombstones), and two *Waihui* (外汇) who are in charge of the accounts. In recent years, the *Bailishihui* has struggled to staff the organization. For example, after being unable for an extended period to find people to take on the *Juzhong* role, in 2007 all of the men in the village between the ages of 45 and 60 were organized into 13 groups, which were to take on the *Juzhong* role in turn. If someone was unable to take on the assigned role, the person had to pay up to 30 yuan (approx. USD 3.9 at the time) per day in penalties.

Amid these challenges, the *Bailishihui* continues to carry out its tasks as before. If a funeral was required for a household which struggled to pay the costs, the *Bailishihui* would go so far as to raise the money, taking the attitude that "if some is too poor to pay for a funeral, all villagers should contribute money and help each other."

With its existing role in funeral ceremonies, the *Bailishihui* became the vehicle through which the villagers' desire to set about rebuilding the temple was channeled. Another significant reason for the *Bailishihui* taking up the role was how it had become seen as the organization "in charge of the deaths of villagers." That is to say, the *Bailishihui* was considered necessary to bring the village together, so that the many deaths due to traffic accidents in the village were not simply treated as "accidental deaths," but were given a greater significance as "deaths within the village community."

However, the *Bailishihui* was clearly not designed to be an organization for rebuilding the *Guandimiao* temple. Hence, a new organization – the Temple Reconstruction Committee – was established in April 2007 with the *Bailishihui*'s backing. Yet, not everyone in Village X was in favor of rebuilding the temple as, for example, there were adherents of Roman Catholicism in the village. Despite the situation, the Reconstruction Committee was tasked to move forward with reconstruction on the basis of consensus among the villagers. What, then, made claiming consensus possible?

Consideration Given to Village Residents in the Temple Reconstruction Committee's Activities

The Committee aimed for thorough resident involvement across all aspects of its initiatives, including construction, appointment of priests, and management. For example, when the Committee was raising funds for the reconstruction, *Bailishihui* executive committee members, including Mr. A and other elderly residents, spent half a month visiting every household in the village. They stood in front of every house explaining the aims of rebuilding the *Guandimiao* temple, asking people to donate even 1 yuan (approx. 0.13 USD) while taking great care to not put pressure on economically struggling families and those who are unenthusiastic about making a donation. By contrast, they also made sure to give due respect for the people who contributed large sums by displaying amounts donated on publicity posters. As a result of such painstaking efforts by the Committee, every single household in the village made some contributions, raising the total amount over 30,000 yuan (approx. 3,900 USD). This was a significant sum of money for a small Chinese farming village, and enough to cover the necessary construction costs.

Reconstruction of the temple began soon after, with the Committee reaching out to villagers with working experience in construction and civil engineering to assist with the project. In response, many villagers volunteered for manual labor such as stacking bricks, or worked for below the market rate on specialized jobs such as plastering. It is fair to say that the rebuilding of the temple was completed by the villagers' own hands.

The reconstruction was completed on October 6, 2007, and a ceremony symbolizing the relocation of the temple from the original site demolished during the Cultural Revolution was solemnly carried out. At the conclusion of the ceremony, fireworks were set off, and the many villagers at the site fell to their knees in worship. For each of the next ten days, theater performances were held to celebrate the completion of the temple reconstruction.

The management and cleaning of the temple is currently undertaken by an elderly woman who lives close by. After the reconstruction, she opened a small variety store selling a wide range of products, including votive offerings. People coming to worship at the temple purchase yellow scripture paper and votive offerings with which to pay their respects; they are effectively, though indirectly, involved in the administration and upkeep of the temple. Since the rebuilding, villagers began to gather at the temple in increasing numbers at significant times such as early morning on the first day of the Chinese New Year, the Lantern Festival, and Guan Yu's birthday, to worship Guan Yu as the guardian deity of the village.

On August 17, 2007, when the reconstruction was still under way, a serious cave-in occurred at a coal mine located around 3 kilometers from the village, leading to the deaths of 172 miners, but fortuitously, not one of the many residents of Village X who worked there was caught up in the accident. This accident was a watershed moment which led the villagers to believe in Guan Yu's power even more.

Reorganization of Village Space and Restoration of Basic Structures of Daily Village Life

Reorganization of Village Space via Reconstruction of the Temple

The hopes and concerns of the villagers who had worked together so successfully to rebuild the temple expanded to include not only fatal traffic accidents, but also the restoration of their living space which had been severed by urban development. The original intention of rebuilding the temple was to treat the deaths of individuals brought about by road projects "as a village matter," and this desire to emphasize the village as a unified entity was also expressed through the physical location of the temple. Until its demolition in 1964 the temple was located in the very center of the village (Figure 7.1). Houses were subsequently built in the original temple area, necessitating the selection of a new site to make reconstruction possible. The traditional *feng shui* belief system in the village held that living in front of the temple is auspicious and living behind it should be avoided, leading villagers to search for a site which was behind every house. This principle was a key driver in settling on the new temple site.[4]

The new location of the temple functioned to clarify the boundaries of Village X. That is, the areas "inside" and "outside" the village became clear, and at the same time it helped to rejuvenate the awareness of the village living space, which had been severed by roads, as one unified whole. As Figure 7.1 shows, the temple was reconstructed on the northernmost edge of the village residential area. By virtue of everyone in the village being involved in rebuilding the temple, villagers developed an awareness that the area to the south of the temple bounded by the major roads (dark shaded areas on the map) was the territory of their village. Put another way, the villagers had positioned the temple as a symbol of the restoration of the village living space. Reunifying the severed spatial order of the village to allow them to live with peace of mind had been a deeply held desire of the villagers.

It also should be noted that in China, rebuilding a temple requires maintaining a delicate balance with local public administration and higher levels of government. In that sense, the reconstruction of the temple can also be interpreted as a case of conflict and negotiation with broader social structural transformation, including the urbanization of rural areas. During the reconstruction project, the villagers took advantage of the unsettled government stance toward temple reconstruction, and doggedly adhered to the position that the rebuilding was an autonomous action by the *ming* ("folk"). This does not mean that the villagers completely avoided the political risk entailed in such a project. What are some of the risk-avoidance strategies that they adopted?

Risk-Mitigation Measures as Part of the **Guandimiao** *Temple Reconstruction*

Anticipating the possibility of being held politically responsible for the rebuilding of the temple, the villagers were prepared to deal with the risk in two ways. The first was having elderly residents take the lead in the reconstruction project.

Their leadership was not limited to fundraising, but their opinions were also reflected in key issues such as site selection. By having elderly villagers such as Mr. A assume leading roles in the project, the villagers hoped that – even if responsibility were to be placed on someone – local government would take no particular action against socially vulnerable people who are "already approaching the next world."

The second strategy was to share the responsibility for the reconstruction project by all village residents. If local government were to take some specific action, it would be undesirable for only Mr. A and the handful of others in leadership roles to be held responsible. To avoid such an outcome, the villagers put in place preemptive measures which they had traditionally used in the event of serious incidents.

For example, the *Bailishihui* would settle funeral accounts in the presence of the mourning family and other stakeholders, and not even the tiniest mistake of 0.1 yuan (approx. 0.013 USD) was permitted. However, there had been one case of a recording error in the past, in which event the treasurer (Waihui) had attempted to compensate for the loss from their own pocket. The *Bailishihui* had responded by refusing to pin responsibility on one individual, instead it decided to cover the loss among all committee members. Their reasoning was that treating what happened as an individual mistake would have a lingering negative impact, making it difficult to find a successor as treasurer in the future.

Such strategies to deal with emergent responsibilities by an organization as a whole, rather than assigning them to specific individuals were also adopted for the temple reconstruction project. Aforementioned Temple Reconstruction Committee, a "folk (*ming*)" organization, was established precisely for that purpose. The tactics used in fundraising and during the construction to encourage participation were all measures to share responsibility among all village residents.

Thanks to the series of these measures, the reconstruction project facilitated a growing awareness among villagers that they are integral member of a community. It goes without saying that the initiative to rebuild the temple alerted higher-level administrative agencies, especially the Xintai City government, and the city government began to pressure the village office to have all the villagers accept the relocation as the city had designated.

At the end of December 2010, each and every villager was required to appear before the village board to give their opinion. Among the villagers present, some had attempted to turn negotiations with the authorities to their advantage by expanding their house to two or three stories, and were thus looking forward to relocation to a housing complex. This is because compensation for relocation would be adjusted proportionately to the floor size of the original residential structure. However, those residents who were looking for relocation changed their minds when faced with the vast majority of villagers refusing to relocate, resulting in the village being unified in its stance against relocation.

What was the eventual fate of Village X? The rural urbanization policy was revised in 2013, and the national government's policy changed its direction from simply demolishing and relocating villages to allowing the possibility of improving existing villages. Village X, which successfully resisted relocation and insisted on rebuilding through preservation, was selected as a National Model Rural Village on December 24, 2019, gaining significant media attention for its village community building.

Conclusions

Through the case of Village X in Shandong Province, which worked to restore the socio-spatial order of the village through the reconstruction of the temple, this chapter has revealed that in times of emergent crises, including those that threaten the lives of villagers and the very existence of the village, what this chapter calls "community *gong*" may rise to the surface. This is not to say that interpersonal relationships – the basis of "personified *gong*" in the discussion of *chaxugeju* – do not matter in forming the character and order of a village community. During "normal" times, such interpersonal relationships continue to play a key role organizing principle in the daily life of modern-day Chinese villages, including Village X. Various studies have reported cases where village elites have fulfilled the demands of the village through their personal networks (Shutō 2003).

Yet, in times of emergency, we witnessed in Village X the formation of "community *gong*," as a distinct organizing principle and normative value, which cannot be reduced to the "personified *gong*" based on interpersonal relationships. The initial catalyst was when the *Bailishihui*, which normally handled village funerals, began to treat the deaths of villagers in traffic accidents as a "village matter." It did so by mobilizing villagers in an initiative outside its organizational aims, namely rebuilding the temple, a place of worship, and a symbol of the village. As this case illustrates, "community *gong*," which in normal times was hidden beneath "personified *gong*" based on interpersonal relationships, was stirred up by the villagers' acute sense of crisis. Once organizations, such as the Temple Reconstruction Committee, rooted in this sense of crisis took center stage in the village, the "community *gong*" began to direct the nature of cooperation among villagers, not seen during normal times, through these organizations. While the power of "community *gong*" prevails, interpersonal relationships stay in the background where such relationships are either considered to be "private" (*si*) matters or mobilized for "community *gong*."

In conclusion, in contrast to those previous studies that emphasize the "personified *gong*" as the governing principle of Chinese villages, this study demonstrates that "community *gong*" does exist, although its existence becomes more apparent and pronounced in the minds of residents in times of crises. In times of significant crises, "community *gong*" may be mobilized, for example, through the creation of unifying symbols and (re)defining their living space, in order to protect and secure their life. It is critical therefore to look beyond the roles of interpersonal relationships that characterize Chinese villages during normal times (however important they are). When the sustained existence of a village is at stake, the village-level, "community *gong*" may well be the last stronghold.

Acknowledgment

The author would like to thank Glen McCabe who translated this article into English. This work was supported by JSPS KAKENHI Grant Number JP20K02108.

Notes

1 The content of the chapter is based on Yan (2013) with substantial revisions.
2 "With individualism, individuals make up organizations in the same way that parts make up the whole. The balance between the parts and the whole produces a concept of equality: since the position of each individual in an organization is the same, one person cannot encroach on the others. It also produces a concept of constitutionality: an organization cannot deny the rights of an individual; it controls individuals merely on the basis of the partial rights they have willingly handed over. Without these concepts, such organizations as these could not exist. However, in Chinese traditional thought, there is no comparable set of ideas, because, for us, there is only egocentrism. Everything worthwhile rests on an ideology in which the self is central" (Fei [1948] 1992, 67).
3 Fieldwork was conducted over seven days in each of February 2009 and February 2011. A number of surnames are used in Village X, such as Wang, Liu, Li, Niu, and Teng. Approximately 90% of the village's population had the surname Wang.
4 According to Mr. A, selecting this particular new location for the temple enabled defining all of the village's residential areas as being incorporated within the boundaries of the temple's territory, rather than only being defined by the new roads.

References

Fei, Xiaotong F. [1948] 1992. *From the Soil: The Foundations of Chinese Society*, translated by G. Hamilton and Z. Wang. Berkeley, CA: University of California Press. (Originally published in Chinese in 1948.)
Fukutake, Tadashi. 1976. *Fukutake Tadashi Chosakushū 9: Chūgoku Nōson Shakai no Kōzō* [The Collected Works of Fukutake Tadashi 9: Structure of Chinese Rural Village Society]. Tokyo: University of Tokyo Press.
Kaino, Michitaka. 1942. *Shina Tochihō Kankō Josetsu: Hokushi Nōson ni Okeru Tochi Shoyūken to Sono Gutaiteki Seikaku* [Introduction to Real Estate Legal Customs in China: Land Ownership Rights and their Specific Characteristics in Agricultural Villages in Northern China]. Tokyo: East Asian Research Institute.
Mizoguchi, Yūzo. 1996. *Kōshi* [Public and Private]. Tokyo: Sanseido.
———. 2001. "Chūgoku Shisōshi ni Okeru Kō to Shi" [Public and Private in the History of Chinese Thought]. In *Kō to Shi no Shisōshi* [The History of Thought Regarding Public and Private], edited by Takeshi Sasaki and Tae-chang Kim, 35–95. Tokyo: University of Tokyo Press.
Shutō, Toshikazu. 2003. *Chūgoku no Jinchi Shakai* [China as a Society of the Rule of Man]. Tokyo: Nihon Keizai Hyōronsha.

Torigoe, Hiroyuki. 1997. *Kankyō Shakaigaku No Riron to Jissen: Seikatsu Kankyō Shugi No Tachiba Kara* [Theory and Practice of Environmental Sociology: Perspectives of Life Environmentalism]. Tokyo: Yūhikaku.

Yan, Meifang. 2010. "Chūgoku Shinnōson Kensetsu ni Miru Kokka to Nōmin no Taiwajōken" [Conditions for Dialogues Between the State and Farmers as Seen in the Construction of New Farming Villages in China]. *Journal of Rural Studies*, 16 (2): 8–19.

———. 2013. "Chūgoku Nōson ni Miru Kyōdōsei to Mura no Kō – Santōshō X-Mura ni Okeru Nōsontoshika o Jirei Toshite" [Cooperativity and Public Village Matters in the Villages of China: From a Case Study of Village X Under Farming Village Urbanization]. *Japanese Sociological Review*, 64 (1): 55–72.

———. 2019. "'Autorō' Teki Kōi no Tadashisa o Sasaeru Chūgoku Seimin no Seitōsei Riron – Tenshin-Shi Busei-Ku X Mura no Danchi-iten o Jirei Toshite" [The Logic of Correctness Among the Chinese People Which Underpins the Legitimacy of "Outlaw"-Type Actions: From A Case Study of Relocation into Housing Complexes in Village X in Wuqing District, Tianjin City]. In *Chugoku no "Mura" o Toinaosu – Ryūdōka Suru Nōson Shakai ni Ikiru Hitobito no Ronri* [Rethinking the "Villages" of China: The Logic of People Living in an Increasingly Fluid Rural Society], edited by Yuko Minami and Meifang Yan, 32–61. Tokyo: Akashi Shoten.

8 Multilayered Commons Space
Dry Riverbed Use in a Local Community in Ibaraki, Japan

Takaaki Isogawa

Riverbeds as a Common-Pool Resources

How should a local community govern its space? This chapter focuses on a particular kind of local space—a dry riverbed, called *kasenjiki* ("river area") in Japanese. *Kasenjiki* refers to a belt-shaped land that is inside the embankments of both banks of a river that has been subjected to hydraulic control work for flood control purposes. It may get flooded during heavy rains or when dams upstream release excess water; otherwise, it is normally used in various forms such as cultivated land, promenades, and sports parks. In Japan, a *kasenjiki* usually exhibits three key characteristics. First, it is legally a public space, administered by the national or local government. Second, it is typically an open and accessible space that is part of the local community's everyday landscape. Third, topographically it is barren land and is not typically seen as highly valuable, leading to a wide range of uses and nonuse depending on local circumstances. Consequently, a *kasenjiki* is a somewhat ambiguous and complexly defined space, similar to *satoyama* or community forest (Takeuchi et al. 2002), which may provide important insights into how inhabitants interact with their local environment.

Kasenjiki can be seen as a kind of common-pool resource, which generally has low *excludability* (ability of the resource owner to prevent or control others from using the resource) and high *subtractability* (the degree to which one's use of a resource reduces the amount available for use by others). One of the central challenges in governing common-pool resources is how to manage excludability, and the issue has been the subject of various theoretical and empirical studies (Suga 2014). In this chapter, I examine *kasenjiki* as a common-pool resource from the perspective of life-environmentalism, with a specific focus on how community actors actually utilize and manage the space. By doing so, I aim to draw policy implications for the governance of commons spaces.

Research Trends in Commons Studies

Research Trends in North America: Building Co-Management Systems

Since the publication of Hardin's "Tragedy of the Commons" model (Hardin 1968), studies of commons have primarily focused on the overuse of common-pool resources. Two contrasting and logical solutions to address the overuse of common-pool resources, derived from this model, are "private ownership and management" and "government/public ownership and management." Nevertheless, many of the subsequent commons studies, particularly those in North America, have explored the effectiveness and sustainability of "community management through community use" as a means to avoid the tragedy of the commons (e.g., Ostrom 2008).

In the North American context, attention has primarily focused on developing institutional conditions to prevent the overuse of commons. This emphasis can be observed, for example, in the Conference on Common Property Resource Management in 1985, which sought to understand why some social groups were successful in managing common resources while others were not (Yokemoto 1998). One significant achievement of commons research was the formulation of eight "design principles" by Ostrom (1990), based on observations of commons where sustainable resource management has been achieved. These principles include clearly defined boundaries, congruence with local conditions, collective-choice arrangements, monitoring, graded sanctions, conflict-resolution mechanisms, minimal recognition of rights to organization, and nested enterprises. Since the development of these design principles, scholarly efforts have largely focused on refining and generalizing the systems of resource management that increase the excludability and reduce the subtractability of common-pool resources (e.g., National Research Council 2002). These studies have achieved some success in the preventing the "tragedy of the commons," such as through the institutionalization of forest resource management schemes in the Global South.

While much effort has been devoted to refining and expanding the institutional design framework, the concept of common-pool resources within the framework appears to have become overly simplified and monolithic. In addition to the two primary characteristics mentioned earlier (excludability and subtractability), important aspects of common-pool resources that can prevent overuse are reduced to properties such as the resource's "renewability, scale, and costs of measurement" (Dietz et al. 2002, 22). In other words, a primary concern of recent research has been how to *improve* the effectiveness of resource co-management systems in order to *maintain* common-pool resources. Within this conceptualization, resource co-management systems undergo critical examination and dynamic reconfiguration, while the resources themselves are assumed to be static and unchanging.

Research Trends in Japan: Reconfiguring Communities

Japan faces a common-pool resource problem that differs from that of North America. In recent years, Japan has been grappling with the underuse, rather than the overuse, of some local common-pool resources due to social changes such as a rapid decline in birthrate and an aging population. Traditionally, Japanese local communities have had mechanisms in place to manage these resources, with clear membership and strict rules for both members and nonmembers. These systems, such as the *iriai* system, have been highly valued by some observers (McKean 2021). However, under the current sociodemographic conditions, there is a growing problem of underuse in local common spaces, often referred to as the "tragedy of the anti-commons" (Heller 1998). This underutilization and undermanagement of local commons spaces are intensified not only by a decrease in the number of users of common-pool resources (e.g., forests), but also because these resources still have legal owners, making it difficult for the government to assume control.

Despite the noticeable difference in the circumstances, Japanese research is similar to studies in North America in that both focus on the effectiveness of resource co-management systems while maintaining a simplified understanding of a common-pool resource itself. In particular, some Japanese scholars and practitioners propose expanding the pool of actors involved in managing the commons in more modern and democratic ways in order to overcome the underuse problems. For example, using such concepts as "collaborative governance" (Inoue 2004) and "environmental governance" (Suga, Mitsumata, and Inoue 2010), they call for the creation of resource management mechanisms that include various stakeholders beyond traditional local actors.

In these lines of thinking, traditional resource management systems are also under scrutiny. Mamiya and Hirokawa (2013, 8) argue, for example, that instead of simply basing environmental policies on the *iriai* system, we must consider how to "open up" traditional *iriai* properties to the broader society. In other words, they argue for a reconfiguration of membership and mechanisms extending beyond traditional local communities. Indeed, some observers such as Miyauchi (2001) propose a new environmental governance scheme that involves updated membership mechanisms incorporating nontraditional and often nonlocal actors who are increasingly active in the realms of common-pool resources.

These emergent arguments cast doubts on the efficacy of the "old community," and argue for its replacement with the "new community" in common-pool resource management. While recognizing the importance of "community," as opposed to "private" or "public" sectors, in governing common-pool resources, these observers advocate for strengthening and expanding "new communities" that resemble modern, civil society organizations, rather than "pre-modern" local community organizations. Simultaneously, there is an assumption that common-pool resources and spaces, such as *satoyama*, should remain relatively static: replenished, but not fundamentally altered. Similar to commons research in North America, Japanese commons research tends to emphasize maintaining common-pool resources with minimal alteration while consciously updating and reconfiguring resource governance institutions, with a particular focus on "new communities."

This chapter questions this popular view and the associated policy approach, suggesting that they may limit the potential for diverse and variable ways of governing common-pool resources. Drawing on a case study of a *kasenjiki* space in Japan and the theoretical perspective of life-environmentalism, the chapter offers an alternative understanding of local space as a common-pool resource. By doing so, it seeks to inform policy approaches that do not solely rely on the reconfiguration and refinement of resource governance institutions.

A Local Community's Interactions with Local Living Space

"Kawabata" Space

Community X, the case study site in this chapter, is a small hamlet comprising approximately 30 households situated in the downstream section of one of the rivers that flow into Lake Kasumigaura, Ibaraki Prefecture. The defining feature of the community is the close proximity of houses built along the river (Figure 8.1). Between the houses and the river, there is a road, embankment structure, and then a *kasenjiki* area spanning about 10 meters in width. The *kasenjiki* serves various purposes for the villagers and is commonly referred to as *kawabata*, which means "by the river" in less technical terms. Residents utilize the *kawabata*

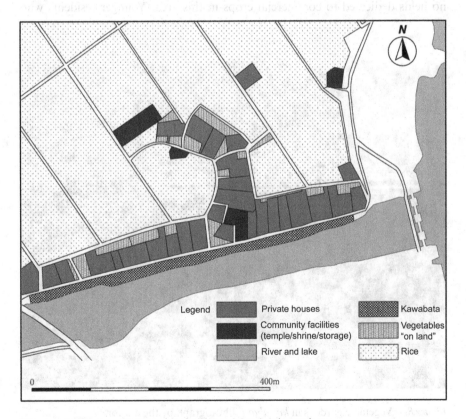

Figure 8.1 Community X and its land use. Map by the author.

for such personal activities as cultivating vegetable gardens, tending to flower beds, setting up simple washing areas, and docking boats. While these uses may appear inconsequential, the patterns of use and accompanying rules elevate this *kawabata* to the status of a common-pool resource within the community, or a commons space.

First, let us examine the vegetable gardens, which represent one of the primary uses of the *kawabata* (Figure 8.2). The majority of individuals tending to these gardens are women in their 60s and 70s. They describe the *kawabata* vegetable gardens as "something that is both reliable and unreliable." They consider them "unreliable" due to the fact that the fields are located outside the embankment and in close proximity to the river, making them susceptible to frequent flooding. However, they also consider them "reliable" because as long as they are not washed away, the soil remains adequately hydrated even during dry periods, allowing plants to stay green. Under such circumstances, the *kawabata* gardens are regarded as a more favorable option than other fields where crops and vegetables may wither.

As these remarks imply, the produce from the *kawabata* fields is not an essential requirement for everyday lives of the community residents. There are no fields dedicated to commercial crops in this area. Younger residents who

Figure 8.2 Vegetable gardens in *kawabata*. Photograph by the author.

work outside the community can conveniently purchase affordable vegetables for their daily consumption on their way home. The women tending to the *kawabata* gardens express that they engage in vegetable cultivation because "grandmas have nothing else to do." However, it is also noteworthy that they frequently share their vegetables and flowers with their neighbors. Residents confirm that the women cultivating these fields enjoy friendly competition among themselves, striving for early timing and high-quality vegetables, which they exchange as part of the harvest.

Community Rules for Kawabata Use

We can see that the *kawabata* is a space that is closely connected with the daily lives of the residents in Community X. However, this does not imply that they have unrestricted freedom to utilize the space as they wish. Instead, there are well-defined rules within the community regarding the allocations of specific areas in the *kawabata*. In Community X, the de facto right to use a particular segment of the *kawabata* belongs to the household adjacent to the land segment. That is, the extension of the housing property boundary serves as the boundary of their corresponding *kawabata* space. Not all *kawabata* segments are actively used by households with the use rights; some segments remain unused at least for the time being. However, the segments that remain unused are still kept free of weeds by the right-holders themselves or with the help of neighboring residents. As a result, the overall condition of the *kawabata* space is well maintained.

It is evident that there is a community-based mechanism to maintain the commons space, which has enabled the sustained use of the space by the community residents to date. However, the mechanism in Community X appears to be far less systematically defined than what one might envision for an efficiently functioning common-pool resource management system, which typically emphasizes high excludability and low subtractability, as often advocated in commons research conducted in North America. As I will demonstrate below, the governance of the *kawabata* space in Community X does not conform to such a well-defined "communal" management system. Furthermore, the *kawabata* space is not sustained by the type of "new community" that some Japanese commons researchers have advocated for. These observations prompt inquiries regarding the exact nature of the commons space within this *kasenjiki* area.

Kawabata: *Public, Communal, and Private at the Same Time?*

A closer examination reveals that the *kawabata* space embodies the aspects of "private," "communal," and "public" simultaneously, but the emphasis on each aspect varies depending on the context. It is in this sense that the space is not governed by a well-defined, strictly "communal" system of management. Below I present five ways in which these aspects are variably emphasized depending on circumstances.

Simultaneously "Private" and "Communal" Space

First and foremost, although the *kawabata* is legitimately considered a common-pool resource, governed by locally shared rules and characterized by communal usage patterns, the actual use is "private" in that each segment is used by a single household. The crucial point here is that the nature of the *kawabata* space, whether it is viewed as "communally" or "privately" controlled, fundamentally depends on the perspective from which it is observed, such as the community or individual users of the land. Any attempt to mold the space into a single category would overlook its true characteristics; it exists as both communal and private simultaneously.

Second, even though the *kawabata* space is used "privately," there are no clear legal boundaries, and they are not delineated by tangible ditches or fences. Moreover, the boundaries are malleable and subject to change based on the relationships between people. For instance, even if one moves to a house facing the *kawabata*, one cannot immediately use the segment of the *kawabata*; the person must first be duly recognized as a member of the village community.

Negotiating with "Government/Public"

Third, the absence of tangible boundaries in the *kawabata* space also reflects its "public" nature. Legally, the river and its *kasenjiki* areas are administered by Ibaraki Prefecture, delegated by the national government. However, negotiations go on between the government and the users of the *kawabata* space. Community members attest that the prefectural office "does not give us a hard time," but they have heard requests such as "do not erect nursery houses or huts," "do not plant trees," and "try to avoid using the *kawabata* (as it is a public space) if possible," which are indirectly communicated during local community meetings. Consequently, even though many residents had previously set up nursery houses in the *kawabata*, most of these structures have now been removed. They also periodically trim trees to maintain a reasonable height. In this manner, local residents are aware that the river, including the *kawabata*, is a "public" space administered by the prefecture. They also understand the administrative logic of "not obstructing the flow functions of the river" and have been more or less responsive to various requests from the government. Therefore, we can conclude that the *kawabata* space, where residents' "private" use based on "communal" rules takes place, also possesses a "public" character.

Function as an Open Space

Fourth, it is worth noting that the "public" nature of the *kawabata* space in Village X extends beyond its ownership and administrative responsibility to its actual use. For instance, the river adjacent to the *kawabata* serves as a popular sport fishing site (e.g., for black bass and crucian carp), attracting many anglers. Naturally, these anglers move around in the *kawabata* space. However, no one in the community view their presence as unauthorized entry or disruptive. In fact, some residents even offer chairs to persistent anglers who remain in one spot to catch fish.

How can we make sense of these behaviors of both anglers and residents in light of the strict communal rules governing the use of the *kawabata* space? The key lies in understanding that the *kawabata* rules are local rules applicable only to legitimate community members. Furthermore, these rules are based on continuous use. Consequently, the local rules of the village cannot be applied to temporary users such as anglers or individuals who come for a walk, and the residents are fully aware of this. There is a shared understanding among residents regarding the appropriate level of "privateness" in utilizing the *kawabata* space and the extent to which "communal" rules apply. This is precisely why the *kawabata* space is used for personal purposes such as vegetable cultivation and drying clothes while simultaneously remaining an open space for the non-local "public."

Figure 8.3 Drying clothes in *kawabata*. Photograph by the author.

Temporal Dynamics

Lastly, it is also important that the current way of using *kawabata* is not rooted in long-standing tradition. In fact, tracing the history of the *kawabata* reveals that its current spatial form only came into existence in 1957 when the *kasenjiki* was established as part of a civil engineering project. During that year, the national government initiated the Agricultural Structural Infrastructure Development Project throughout the entire village. Prior to this project, most of the river shores that now make up the *kawabata* were actually used as residential land and rice fields.

As part of the project, the village's roads were realigned along the river, and embankments were constructed. Consequently, the original residential land and rice fields had to be relocated further inland. This transformation created a narrow strip of "gap" area between the embankment and the river, which eventually became the present-day *kawabata* space. In essence, the *kawabata* in the village is a relatively recent development that emerged through the residents' land utilization, their negotiations with formal administrators, and various other circumstances, ultimately leading to the establishment of local rules for its use.

Toward a Deeper Understanding of Commons Space

Kawabata, as a common-pool resource in Community X, represents a space where elements of "public," "communal," and "private" coexist and overlap, with their emphasis and tones shifting based on the specific conditions and perspectives of the actors involved. Furthermore, the nature of the space is multilayered and subject to change, both in the present and over time. Recognizing this complexity is crucial when engaging in policy discussions, as they should take into account and reflect the dynamic characteristics of the space.

To explore this further, let us revisit a theoretical debate in commons research. Hiroyuki Torigoe (2010, 62) highlights two main approaches to understanding commons. The first approach sees "public," "communal," and "private" as three independent balloons within a closed space. According to this view, when one balloon expands, the others must shrink, suggesting a relationship of rivalry. The second approach views "public," "communal," and "private" as three flat rice crackers on a plate (like a Venn diagram), allowing for layering of these elements. The former are referred to as "discrete-type commons," and the latter as "layered-type commons."

Applying this classification, we can observe that discussions on commons in both North America, which emphasizes strengthening and refining co-management systems, and Japan, which focuses on reconfiguring institutional arrangements through modern and democratic procedures, are based on the assumption of "discrete-type commons." In this approach, the recommendation is to replace the outdated or declining "old" communal management system with a "new" one, expanding the "new" communal management while scaling back the "private" and "public" spheres.

In contrast, the observations of the *kawabata* space of Community X align better with the concept of "layered-type commons." In this understanding, we do not see that "public," "communal," and "private" as fixed characters of a given space; instead, they are layered on top of each other. It follows that one of the characteristics may appear dominant in a particular context, while other "latent" characteristics remain less visible at that moment. However, if circumstances change, those other characteristics may become more pronounced.

Conclusions

The case study of how local people understand and use *kasenjiki* in Community X has revealed the complex nature of a common-pool resource that is simultaneously "public," "communal," and "private." Moreover, those different aspects are not merely overlapping, but each aspect may manifest itself in varying intensity depending on the context. Thus, *kasenjiki* can be seen as a dynamically variable space.

Typically, policy discussions regarding common-pool resources focus on how to counterbalance the expansion of the "private" and "public" spheres through the promotion of "communal" management. The aim is often to replace dysfunctional "old" communal resource management systems with more functional ones and to strengthen and refine them further. However, based on the findings of this study, policy approaches that assume fixed characteristics in a given local space must be critically reconsidered.

Admittedly, the "layered-type commons" can be relatively unstable. The current characteristics of the space ("public," "communal," or "private") are contingent upon the ways in which people have interacted and currently interact with the space under various conditions. Consequently, these characteristics may well be transitional or temporary. Nevertheless, if this is how people engage with the local environment, policy discussions should embrace and acknowledge the multilayered and variable nature of the commons space, rather than attempting to oversimplify or ignore it. Adopting such a policy perspective will lead to more sustainable engagement of residents with their local environment in the long run.

References

Dietz, T., N. Dolsak, E. Ostrom, and P. C. Stem, 2002, "The Drama of the Commons", Committee on the Human Dimensions of Global Change, *The Drama of Commons*, 3–35. Washington D.C.: National Academy Press.

Hardin, Garrett, 1968, "The Tragedy of the Commons," *Science* 162: 1243–8.

Heller, Michael A. 1998 "The Tragedy of Anticommons: Property in the Transition from Marx to Markets," *Harvard Law Review* 111 (621):622–88.

Inoue, Makoto. 2004. *Komonzu no Shisō o Motomete* [Seeking Ideas of the Commons], Tokyo: Iwanami Shoten.

Mamiya, Yōsuke, and Yūji Hirokawa. 2013. "Komonzu Kenkyū no Kiseki to Kadai" [The Trajectory and Challenges of Commons Research]. *Komonzu to Kōkyō Kūkan* [Commons and Public Areas], edited by Yosuke Mamiya and Yuji Hirokawa, 1–18, Kyoto: Showado Publishing.

McKean, Margaret A. 2021. "Management of Traditional Common Lands (Iriaichi) in Japan." In *Making the Commons Work*, edited by Daniel W. Bromley, D. Feeny, P. Peters, J. L. Gilles, R. J. Oakerson, C. F. Runge, and J. T. Thomson, 63–98.

Miyauchi, Taisuke. 2001. "Kankyō Jichi no Shikumi Zukuri–Seitōsei o Kuminaosu" [Environmental Autonomy and Restructuring of Legitimacy]. *Kankyō Shakaigaku Kenkyū* [Journal of Environmental Sociology], 7: 56–71.

National Research Council. 2002. *The Drama of the Commons*, edited by Elinor Ostrom, Thomas Dietz, Nives Dolšak, Paul C. Stern, Susan Stonich, and Elke U. Weber. Washington, DC: The National Academies Press. https://doi.org/10.17226/10287

Ostrom, Elinor. 1990. *Governing the Commons: The Evolution of Institutions for Collective Action*. New York: Cambridge University Press.

———. 2008. "Institutions and the Environment." *Economic Affairs* 28 (3): 24–31.

Suga, Yutaka. 2014. "Gabanansu Jidai no Komonzuron–Shakaiteki Jakusha o Hōkatsusuru Shakai Seido no Kōchiku" [Commons Theory in the Governance Era: Construction of a Social Institution Inclusive of the Socially Vulnerable]. In *Ekoroji to Komonzu* [Ecology and the Commons], edited by Gaku Mitsumata, 233–52. Kyoto: Koyoshobō.

Suga, Yutaka, Gaku Mitsumata, and Makoto Inoue. 2010. "Gurōbaru Jidai no Naka no Rōkaru・Komonzu-ron" [Local Commons Theory in the Middle of the Global Era]. In *Rōkaru Komonzu no Kanōsei* [The Possibilities of Local Commons], edited by Gaku Mitsumata, Yukata Suga, and Makoto Inoue, 1–8. Kyoto: Minerva Publishing.

Takeuchi, Kazuhiko, et al. eds. 2002. *Satoyama: The Traditional Rural Landscape of Japan*. Tokyo: Springer-Verlag.

Torigoe, Hiroyuki. 2010. "Shohyō Murota Takeshi Hencho 'Gurōbaru Jidai no Rōkaru Komonzu' (Minerva Shobō, 2009 Nen)" [Book Review: Local Commons in Globalization (2009), edited by Takeshi Murota, Minerva Publishing]. *Zaisei to Kōkyō Seisaku* [Public Finance and Public Policy] 32 (2): 56–62.

Yokemoto, Masafumi. 1998. "Shizen Shigen no Keizai Bunseki to 'Komonzu-ron no Shin Tenkai'" [Economic Analysis of Natural Resources and the Theory of Commons]. *Hitotsubashi Kenkyū* [Hitotsubashi Journal of Social Science], 22 (4): 115–28.

Part III
Living with Disasters

9   Why Do Victims of the Tsunami Return to the Coast?*

Kyoko Ueda and Hiroyuki Torigoe

Even after a natural disaster of tremendous scale occurs, some victims attempt to remain or later return "home" while inviting the risk of experiencing further catastrophe. As reported worldwide, the Great Tōhoku Earthquake brought about a large-scale tsunami that inflicted devastating damage on the inhabitants of fishing villages along the Pacific coast of the Tōhoku area. The Sanriku region, situated in a seismically active area, has repeatedly incurred serious damage from tsunami. Hence even the tsunami that occurred on 11 March, 2011 was actually not an unprecedented event in history. So why do people opt to continuously live where they are prone to natural disasters instead of living at a distance from the coastline? Especially for those who have just experienced the tsunami, what motivates them to make the decision to go back to the coast?

The purpose of this chapter is to clarify why people continuously live in places where a specific natural disaster comes with such apparent frequency. In particular, this chapter refers to the case of a coastal area called Sanriku where tsunami have recurrently hit at least every 40 years for the last 115 years, including the one caused by the Great Tōhoku Earthquake. And this most recent tsunami is also said to fit into a 1000-year tsunami cycle (at least from 6,000 years ago) according to one geographer's analysis of the geologic strata found south of Sanriku.[1]

Located in the Tōhoku region (in the north-eastern part of Japan), the Sanriku region stretches along the Pacific Ocean coast. Embracing a sea rich region with a large variety and rich catch of fish, its coastline, named the Rias coast, maintains a geographic vulnerability, which induces tsunami and amplifies their force of it. This geographic character of the coastline provides both rich fishing ground and tsunami, historically. Hence, not a small number of villages along the Sanriku coast have already lived through the devastating damages of tsunami repeatedly.

In order to examine why people who have just experienced a tsunami may try to go back to the Sanriku coast, this chapter focuses on the case of one small community which lost 44 out of 52 homes, and more traumatically four

---

* This chapter was originally published as Ueda, Kyoko, and Hiroyuki Torigoe. 2012. "Why Do Victims of the Tsunami Return to the Coast?" *International Journal of Japanese Sociology*, 21: 21–29, and is used with permission.

villagers, to this most recent and utterly devastating tsunami of 11 March. An attempt will be made to understand the strategies of the inhabitants of coastal villages who, assumingly, somehow find the means to, or manners for, culturally interpreting and "domesticating" marine catastrophes into their community history. Upon this premise, this chapter starts with the "unusual day" the tsunami arrived and follows this damaged community's decision-making process, leading toward a return to their "home." Simultaneously, their "usual" days, which were suddenly disrupted, will be inevitably referred to in pursuing their reasoning to return to their native home.

Sociological (and Anthropological) Approaches to Natural Disasters

In describing a community that tries to return to its original place near the sea despite having been devastated by a tsunami, a series of works by Yaichirō Yamaguchi cannot be ignored. Yamaguchi, while listening to every detail of the people's ordinary days, walked through the villages of the Sanriku coast to research the damage, the evacuation process, and the resettlement patterns. In exploring the reasoning for their return to the coast, he describes the process of a community's resettlement back to the seashore as follows. Although the villagers have at first resettled on higher ground after suffering the destructive force of a tsunami, people gradually start to move back to the original places one by one. For example, newcomers begin to settle back down in the formerly populated areas, which are much closer to the sea than the newly settled village areas that have been created in order to avoid future tsunami. However, when newcomers, who historically have a relatively higher mobility and lower income (Yamaguchi 2011, 208), start to yield a richer catch of fish because of their close proximity to the sea, then tsunami victims, who were older inhabitants, once again start to move down to the coast, feeling that they are being deprived of the resources of their livelihoods (Yamaguchi 2011, 211).

While Yamaguchi acknowledges that the main reason for coming down to the seashore is such an economic motive, he does not attribute the survivors' return solely to the relative deprivation of their livelihood. He explains that the reason to come down to the original place near the sea derives from a worship of ancestors that are believed to remain at the ground of the original house, paddy field, or graveyard. Thus, the older villagers get, the more they feel like going back home because of their attachment to, not merely their homelands, but more precisely, their ancestors (Yamaguchi 2011, 210). With his detailed understanding of the everyday life of coastal villages and profound consideration of their culture, he had to admit a sort of "irrationality" in peoples' decision-making in which they return to the seashore after the torrent of tsunami. His stance surely comes from his experiencing, seeing, and hearing about the great sacrifices made because of the recurrent attacks of tsunami.

Whereas Yamaguchi could not deny the people's tenacity in explaining the reasoning of their resettlement back in the disaster area, studies on natural disaster by anthropologists and sociologists attempt to reveal some kind of "rationality" within the victims' decision-making. This contrastive approach is due to the radical anthropological and sociological understanding of natural disasters:

> In the face or threat of disruption, as people attempt to prepare, construct, recover or reconstruct, how they adjust to the actual or potential calamity either recants or reinvents their cultural system. Disaster exposes the way in which people construct or frame their peril (including the denial of it), the way they perceive their environment and their subsistence, and the ways they invent explanation, constitute their morality, and project their continuity and promise into the future.
> (Hoffman & Oliver-Smith 2002, 6)

Such an approach allows some room to seek out the rationality within the victims' or survivors' strategic behavior throughout the unfolding process of a disaster, from the day it occurred until normality returns. By adopting this viewpoint, this chapter attempts to understand the rationality in tsunami victims' returning to coastal areas by describing the detailed process of survivors' decision-making. In addition, an exploration will be conducted on how tsunami are socially embedded in their society.

The Sanriku Region and Tsunami

The History of Tsunami in Sanriku

One episode relating to the degree to which people in the coastal area of Sanriku are "familiar" with tsunami is that they could tell to which point the water came "the last time." Hence many of those who died due to the attack of the tsunami on 11 March, 2011 by the Great Tōhoku Earthquake were inhabitants whose houses were placed above or far away from the waterline of the last large-scale tsunami, that is, the Showa-Sanriku-Otsunami in 1933.[2] As mentioned above, the Sanriku region has experienced periodic large-scale tsunami approximately once every 40 years for the last 115 years: Meiji-Sanriku-Otsunami in 1896 with 21,953 casualties and 7,274 houses lost; Showa-Sanriku-Otsunami in 1933 with 3,064 casualties and 6,067 houses lost; and the Great Tōhoku Earthquake in 2011 with 19,846 casualties and 295,018 houses lost (Kokuritsu-Tenmondai 2011, 720–753).

However, the three cases cited above are only the large-scale tsunami. If we include smaller-scale ones, the residents of the Sanriku district experienced tsunami with much greater frequency. Older historical records note large-scale tsunami disasters: in the year 869, the Jokan Tsunami, M8.3, with more than 1,000 casualties; in 1611 the Keicho Tsunami, M8.1, with more than 3,000 casualties; in 1677 the Enpo Tsunami, M8, with 12 casualties with a loss of 6,060 houses; and the Ansei Tsunami in 1856, M8. Thus, on average, the Sanriku region has coexisted with tsunami on average every 46 years[3] for more than 1,000 years.

Therefore in the coastal villages of Sanriku, numerous stone monuments have historically been erected with written warnings, such as "As soon as an earthquake occurs, be careful of tsunami." Interestingly, the places where these stone monuments stood, in most cases, marked the farthest intrusion of water inland of the Showa-Sanriku-Otsunami or the Meiji-Sanriku-Otsunami (Kawashima 2011); however, the water produced by the most recent and probably the worst tsunami, that of March 2011, exceeded far beyond the point of many of these monuments.

Despite such recurrences of destructive tsunami, the history of villages in Sanriku is quite old. This means that people have continuously lived along the coast of Sanriku for more than 1,000 years while coping with tsunami, instead of moving away. In addition, most of their villages have repeatedly engaged in a "back-and-forth" resettlement; at times moving a bit higher, further away from the shore after experiencing a large-scale tsunami, and then back toward the seashore after a lapse of time.

A hamlet called Moune, being the smallest unit of a rural village community referred to as *sonraku* or *shuraku* in Japanese, is one such fishing village that decided to move back to the seashore after enduring the tsunami of the Great Tōhoku Earthquake. The data discussed in this chapter from the next section is based upon field research conducted in Moune. Both their manner to evacuate at the onset of the tsunami and the strategies employed for their return to the coast will be described while exploring the reasoning to do so.

What Happened on 11 March in Moune

Moune, facing the small and quiet inner bay of the Karakuwa peninsula, is a small community that belongs to the well-known larger port town of Kesennuma City having a population of about 70,450, and 25,595 households[4] (Figure 9.1). As the largest landing harbor in the Tōhoku district, the port of Kesennuma is not merely a local port but accommodates tuna and bonito fishing vessels from all around Japan. Compared to the city center of Kesennuma, Moune is located in typical Japanese countryside surrounded by gentle hills, which makes it a relatively isolated community geographically.

The damage in Kesennuma City as a whole, caused by the 2011 tsunami, was quite serious with 1,390 casualties (including those who are missing), a loss of 15,103 houses, and a destruction of more than 90 percent of boats moored within the city. The height of the tsunami, which assaulted Kesennuma port, has been reported as exceeding more than 10 m.[5] This destructive wave also destroyed 21 huge fuel tanks (out of a total of 23 equipped at the port) that resulted in 12,810 kL of oil spilled and a conflagration that lasted for 4 days in the center of the city.

Likewise, Moune also experienced a devastating tidal wave with a height of more than 11 m losing four of its inhabitants, 44 houses (out of 52), and almost all of its boats (25 out of 27) moored within the Moune Bay. After the M9 earthquake struck with its intensity of 6,[6] it took around 30 min until the first

Figure 9.1 Moune community in Kesennuma City, Miyagi Prefecture. Map by Daisaku Yamamoto.

wave arrived at Moune Bay. During this half-hour period, people in Moune were quite certain that a large-scale tsunami would come because of the fierce quake. They say that, first, the bottom of the sea was exposed to a wide extent, and just beyond, the surface of the sea rose up rapidly. Most of them climbed uphill with their family or community members while checking houses to see if older people or children still remained behind to rescue them. Some moved cars to higher places so as to avoid the tsunami assault.

People who were by the sea or on boats quickly left the bay to head out toward the ocean, in other words to the direction from where the tsunami was approaching, because for ships it is said to be safer as long as they stayed on the open surface of the sea. For fishermen in Kesennuma, it was common knowledge that if ships are larger than 4–5 t with adequate engine power, it is better to move out toward sea in order to better protect the ships.[7] This is called *oki-dashi* and is the recommended manner by which to save their ships in case of tsunami. Since the fishing boats are indispensable tools for their livelihood, *oki-dashi* has been repeated every time a tsunami enters Moune Bay, regardless of its predicted size. This time, there were four ships that were large enough to execute *oki-dashi* moored within the local bay. Although three of them attempted *oki-dashi*, one failed because of the power and speed of the tsunami. This ship itself was in danger and so the pilot abandoned it and swam to the nearest island, Oshima. He reappeared at home safely three days later.

Up Until the Decision Was Made to Return to the Coastal Area

Hence the people of Moune, except for the four who died, somehow managed to escape the tsunami. And 8 houses out of 52 survived. The survivors in the community of Moune, however, became isolated because all the ships disappeared after the tsunami hit and the road connecting it to the city of Kesennuma and to the center of Karakuwa was disrupted. With the disruption of electricity and tap water, people within the community gathered together and survived with water from a nearby stream, along with firewood and food stocked in houses safely placed uphill. After two days, one ship, which had undertaken *oki-dashi* and successfully escaped the tsunami, came back to Moune Bay and was warmly welcomed by all the survivors.

As soon as this ship arrived at the shore, it ferried people one after another to the elementary school in Karakuwa, which was their designated place of refuge in an emergency. They stayed there for four months until those who lost their homes were moved to temporary emergency housing. The remaining eight households, whose houses survived, stayed in Moune, but at the cost of being separated from most of the community members.

During the four months following the disaster, they had to decide if they should return to their homeland together as a community or just to allow each household to decide on its own without any community consensus. Their strategies to move back to the coast, however, had already started in the week following the tsunami. They talked about the community's resettlement over candlelight as electricity was not soon restored. Twelve days after the earthquake, on 24 March, a representative, whose household was the earliest settler in the community, visited the local government office to ask about resettlement back to their home. Despite his earnest inquiry, an official told him that the matter could not be dealt with because his claim was made individually. In order to persuade the official of the local government, an organization was needed which could publicly represent the consensus of victims in Moune. Thus, the people of Moune set up an organization (the *Mounechiku Bōsai Syudan-iten Sokushin-jigyō Kisei Dōmeikai*), which qualified them to officially petition for resettlement. Consequently, 30 out of 44 households, those whose houses were swallowed up by the tsunami, joined together in the form of this organization in order to return to their coastal community.

In less than a week after losing their homes, boats, and most of their belongings, they had discussed returning to the coast. Two months later, they even traveled 450 km away to Nagaoka City to meet with victims of the Chūetsu Earthquake (2004) who had successfully managed to resettle their community back in its original place.

What compels them to go back to the seashore where the destructive tsunami deprived them of almost everything? It is true that they are a fishing community, but only three households were exclusively engaged in fishing or aquaculture (oyster and scallop) before the earthquake. The rest of the households owned small boats known as *iso-bune* for collecting abalone, sea urchin eggs, and clams, as well as for harvesting seaweed, but their catch only occupied a small part of their livelihood, or they sometimes did not sell anything at all. Fishing was no longer the main livelihood for most of the households whose heads were old enough to live on their pensions. In addition, although many community members engaged in oyster farming, run by other households, the tsunami swallowed up all the oyster beds and all the boats except for three. Hence their economic motives were not necessarily decisive in their reasoning to reenter the coastal areas. In the next section, the experiences of living by the coast for people in Moune will be explored by looking at their practices to address both the vulnerability and the fertility of the sea in their community history.

What "Going Back to the Seashore" Means for Them

Strategies to Live by the Coast

People in Moune equally state, "we were not at all prepared for the tsunami, and this is how it went." In fact, the present community members should have known directly or from hearsay experience that large-scale tsunami have hit Moune three different times: the Meiji-Sanriku-Otsunami, M8.1, in 1896 with a 3.3-m wave height, 13 casualties and 19 houses lost; the Showa-SanrikuOtsunami, M8.1, with a 2.15-m wave height and 24 houses flooded; the Chile Earthquake-Tsunami in 1960 without any loss of life or homes but with a loss of all the ships and oyster beds as should be equally noted for the two former cases. Actually, there is a family whose house was destroyed by tsunami three times: in 1896, 1933, and 2011.

In addition, the Chile Earthquake-Tsunami of 2010, a year before the Great Tōhoku Earthquake, also completely reduced all the oyster beds hung within Moune Bay. The present style of raising oysters and scallops was established in the 1920s and was widely introduced to Moune Bay after the Showa-Sanriku-Otsunami. Until then, piscary or fishing rights did not exist, and oysters were raised not in suspension style hung from the rafts as they are today, but were raised with the spat hung from stakes that were fixed at the shorefront.

Whereas the transition to the newer style of oyster aquaculture dramatically increased the productivity, the rafts for oyster spats are highly fragile to the shock of tsunami. As almost all of the households engaged in this aquaculture, Moune Bay gradually became crowded with the wooden rafts used to raise oysters; they covered the sea surface almost to the outer sea. Therefore, the oyster's growth was exposed to competition among the community members. They say, "Four sets of rafts were enough to somehow feed the family." This does not mean that people in Moune were exclusively engaged in oyster aquaculture since the 1920s; they fished, cultivated the small tracts of land, collected mushrooms from the mountains, and earned a living. However, they increasingly came to depend on oyster farming as central to their livelihood. Simultaneously, their adoption of this more productive way of aquaculture led to a promotion of vulnerability to tsunami because it uses a suspension type of oyster farming that is highly fragile to the impact of tsunami. In a way, the more they gained from the sea, the more vulnerable their community became to calamities. Therefore, they lost all the oysters with every tsunami since the adoption of the aquaculture with rafts. Hence people in Moune have restored the oyster beds repeatedly. On the surface of Moune Bay, there are already newly crafted oyster rafts afloat just four months since the earthquake.[8] Having said this, should we understand that people in Moune are not really prepared for these disasters at all?

Moune residents became increasingly dependent on oyster farming, whereas aquaculture had been continued as a kind of risk diversification of their livelihood. Before wage labor in town had become popular, the typical division of labor carried out by families in Moune was divided as follows: fathers (up into their 40s or 50s) engaged in tuna or bonito fishing on long ocean cruises, being absent from the community for several months; and grandfathers and grandmothers (sometimes along with grandchildren) work for aquaculture inside of the bay. When a father got older, the long cruise fishing was taken over by his eldest son. Then the father retired from the tuna or bonito fishing and joined in the aquaculture within the bay. Thus, people in Moune were prepared for a poor catch, which might happen both within the bay and outside of the bay, by diversifying the risk. It is probable that they would follow this way after resettling back in Moune, by continuing fishing or aquaculture within the bay, and receiving a pension instead of deep-sea fishing. Hence, in addition to stabilizing the harvest of oysters, they split the risk of a poor catch between inside and outside of the bay.

What Have the People of Moune Learned in Their Repeated Experience of Catastrophes?

In a way, the people of Moune know how vulnerable oyster farming is to tsunami; therefore, they diversify the risk by catching fish from both inside the bay and out at sea. They also know that when a tsunami comes the inner bay becomes highly dangerous, and, in contrast, the outer sea is safe as long as they are on ships. In other words, they know the various dimensions of the sea, which greatly condition people's living by the coast.

The people of Moune also know that after being hit by a tsunami the sea becomes purified and rich in fish. A man who was the first to restart his oyster farming after the attack of the tsunami said,

> Just after the tsunami came, I felt at a loss. However, the attack of the tsunami also reminded me of the memory of purity and richness of the sea just after the attack of Showa-Sanriku-Otsunami. Honestly, I thought that this time the tsunami could be too huge to revive and purify the sea. I thought all the creatures under the sea could be totally 'dead' because all kinds of fish, starfish, sea urchins and even wharf roach disappeared

from Moune Bay for two months. However, my grandson said to me one day, 'grandpa, fish are swimming in the sea.' So I came down to the bay and found the sea was filled and crowded with fish in a highly transparent water."[9]

In fact, the people engaged in oyster farming after 3/11 say that the oyster spat grows quicker than ever before because of the purity of the sea after this tsunami and that they ran out of hands to hang them. They now grow so fast even while they are stored in the seawater. Thus, in people's memory, Moune Bay revived after the destructive attack of the tsunami.

While adapting to the multidimensions of the sea, the people of Moune have also established a manner by which to face the tragedy brought about by maritime accidents or marine disasters. In case of the loss of people at sea, the people in Moune repeated a ritual called *hama-barai* or *ura-barai* in the local vernacular, in order to comfort the deceased and feed them fresh water because they are supposed to have died while taking in seawater. Interestingly, at the beach of Moune Bay, they summon all the spirits that are believed to have died at sea and comfort them time and again. Every time any community member has an accident at sea or when an anonymous dead person drifts ashore in the local bay, they cannot board the ships or fish without first conducting the *hama-barai* because their local bay, they believe, becomes "unclean." Therefore, they believe that the sea has to be purified, in a symbolic sense, otherwise, all the ships which leave Moune Bay cannot enjoy the abundant fish; moreover, another tragedy might succeed. After the tsunami of 3/11, they have done this *hama-barai* "as usual." The unusual part this time, however, was that it was performed together with all the neighboring communities at one port because most of the individual local ports or beaches were filled with rubble and too damaged to approach; and it took 100 days from 3/11 until they performed this ritual. This was in consideration for the survivors who lost their families in addition to their homes and were not yet ready to accept the sudden death of their closest relatives. In Moune, however, although their original place in the community became partly flooded with seawater around the time of high tide because sinking of the land caused soil subsidence brought about by the earthquake, they have done another traditional ritual, called *misoka-bon*, to comfort their ancestors in Moune Bay. Although they had to wear rain boots and get together from the shelters in several locations, they repeated the rite as usual for the deceased who will also join their ancestors.

Somehow the uncleanliness, brought about by the tremendous number of casualties on 3/11, was symbolically taken away from all of the beaches and ports by the survivors of the communities. Thus, they become ready, in a symbolic sense, to get on the ships again and fish both inside and outside of each local bay.

In a way, this practice can be interpreted as an unusual event; a catastrophe that was far beyond their imagination became a more familiar "unusual event," and what we might even say was "usual." The damage from the tsunami on

3/11 was surely devastating for them. Nevertheless, they somehow knew how to calm or comfort the deceased at sea, and to take away the uncleanliness of the sea in a spiritual sense. From their long history as a coastal community, they somehow knew how to deal with a death, without having an actual corpse, as many disappeared into the sea, and also coped with anonymous corpses, which drifted into the local bays. Only after this ritual was done could they enjoy the fertility of the sea again. This would speak to the resilience of the people in Moune that was realized in the process of returning to their village by the seashore.[10]

Inseparable Hazard and Livelihood: "Monster and Mother"

As the Moune community indicates, people know that life near the coast is inevitably entwined with both the severity and fertility of the sea. In other words, what people in Moune know is that they cannot have one without the other. Both sides of the sea have conditioned the life of Moune people and that is what they have adapted to. Because of their closeness to the sea, the fertility they have enjoyed and the vulnerability as a coastal community are inseparable for them, like two sides of the same coin. They know that both are born from the same sea. In Hoffman's words, the people of Moune understand "the monster and the mother" as one integrated personality (Hoffman & Oliver-Smith 2002, 114).

To prevent the damage of disasters and to maintain a resilience against disasters are not equal. Surely, the people of the Moune community might have done nothing more than escape the onslaught of a tsunami of tremendous scale. Nevertheless, just as their vulnerability came to light, their resilience to reconstruct a coastal community has also gradually become apparent. They did not know how to save all members of the community, their homes, nor their ships. Whereas they did know how they should adapt to the inherent instability of their coastal community and how they should revive the community. This process would unfold only after having an understanding of the way in which they needed to adapt to their local bay. As a coastal community they were ready to accept their vulnerability, even immediately after the destruction of the tsunami, as it was the condition in which to enjoy the fertility of the sea as their ancestors had done throughout history.

Notes

1 From the strata found in Kesennuma, north of Sanriku by Hirakawa. Yomiuri Online, 22 August 2011. http://www.yomiuri.co.jp/science/news/20110821-OYT1T00511.htm, accessed on 26 January 2012.
2 This is according to interviews with inhabitants conducted in the community named Syuku, next to Moune in Kesennuma City. Also, research by the Japanese Ministry of Land, Infrastructure, Transport and Tourism shows that the death rate in the plain area, caused by the tsunami, was higher than that of the coastal areas as reported by Yomiuri-Shimbun on 5 October 2011.

3 Based on the article by the Cabinet Office of Japan. http://www.bousai.go.jp/jishin/chubou/kyoukun/4/shiryo4-1.pdf, accessed on 26 January 2012.
4 Based on the statistics of Kesennuma City as of October 2011.
5 Based on the statistics of Nihon Kishō Kyōkai, http://www.jwa.or.jp/static/topics/20110329/touhokujishin110329.pdf, accessed on 26 January 2012.
6 This index is based on the Japanese seismic scale of seven stages.
7 According to an interview with the head of the Karakuwa branch on 12 July 2011, the local fisheries cooperative association (Miyagi-ken Gyogyō Kyōdō Kumiai Karakuwa-shisho) does not officially recommend *oki-dashi* because its success depends on the environmental conditions and the various characteristics of the ships.
8 This could happen with support and donations from volunteers and from domestic and foreign companies.
9 From an interview in Moune on 13 September 2011.
10 Editors' note: To be precise, the tsunami-afflicted households of Moune did not return to their original locations of their houses. Rather, a group of 24 households collectively resettled in a higher ground near their original housing sites between 2015 and 2016. This collective resettlement was one of the speediest resettlement projects within Kesennuma City. The original village area subsided and became a salt marsh. A local NGO has been working to preserve this emergent natural environment.

References

Hoffman, Susanna M., and Anthony Oliver-Smith, eds. 2002. *Catastrophe and Culture*. Santa Fe: School of American Research Press.

Kawashima, Shūichi. 2011. "Shinsuisen ni Matsurareru Mono" [Things Enshrined in the Flood Lines]. In *Kikan Tohokugaku*, 29, edited by Tohoku Bunka Kenkyu Centre, and Tohoku Geijyutsu Kouka Daigaku, 27–37. Tokyo: Kashiwa-Shobō.

Kokuritsu, Tenmondai, ed. 2011. *Rika-Nenpyo Heisei-24* [Chronological Scientific Tables 2012]. Tokyo: Maruzen-Syuppan.

Yamaguchi, Yaichirō. 2011. *Tsunami to Mura* [Tsunami and Villages]. Tokyo: Miyai-shoten.

10 The Roots of Resilience

Forest Commons and the Cultivation and Disappearance of Livelihood Security in a Nuclear Disaster–Afflicted Community

Hiroyuki Kaneko

Life in Kawauchi, a small village of 2,000 residents situated in the mountainous forests of Fukushima Prefecture, has until recently centered on forestry and subsistence agricultural activities. Forests occupy 17,000 ha (85%) of the village's total area, while agricultural land is limited to only 970 ha (5%). Until the 1950s, residents earned their living by harvesting forest resources to be sold as timber and charcoal and growing crops in the narrow fields of the village for self-consumption. Since the 1960s, however, nuclear and thermal power plants began to be built one after another throughout neighboring municipalities along the northeastern coast, and the livelihoods of Kawauchi became increasingly integrated into the energy industry.

Although life in the mountainous village of Kawauchi had changed dramatically over the previous half century, few could have imagined the catastrophic transformations of 2011. A series of hydrogen explosions at the Fukushima Daiichi Nuclear Power Plant, located only 30 kilometers from the village, triggered by the great earthquake and tsunami of March 11, led the Japanese government to issue a comprehensive and mandatory evacuation of the entire village on March 15 (Figure 10.1). Fortunately, the village was not in the path of prevailing winds at the time of nuclear fallout, and the level of radioactive contamination in Kawauchi was not as severe as that of towns and villages to the north. Eventually all restrictions on Kawauchi residence were lifted by June 2016 after the decontamination of farmland and residential areas through the removal of topsoil, pruning, and intensive cleaning.

Nevertheless, the vast forests of the village, which cover nearly 90% of total area of Kawauchi, have been left untouched to date. Opting not to decontaminate Kawauchi's vast forests appears economically rational because forestry in the area, as in much of Japan, has become increasingly unprofitable, and the costs of decontaminating these areas are tremendous. However, as this chapter will demonstrate, this decision represents a continued and increasingly complete slashing of the ties that have connected forest and community in Kawauchi for centuries. As I have shown elsewhere (Kaneko 2017), this decision appears rational from a narrow economic calculus, but is highly problematic when we recognize that the forests of Kawauchi have played prominent roles in the

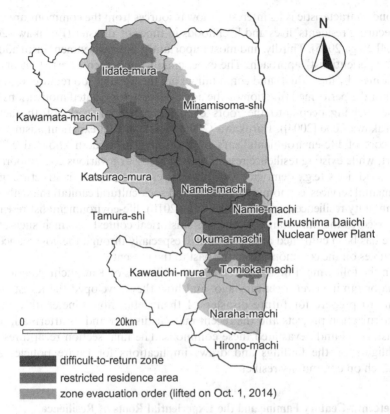

Figure 10.1 Evacuation designations in the area around Fukushima Daiichi as of September 2014. Created by the author based on Ministry of Economy, Trade and Industry (2014).

subsistence and sociality of village life. However, by focusing on forest use in the present day, my previously published research has also under-emphasized the ways in which these common spaces have supported community resilience historically, and how past national government forest policies had problematically reduced community resilience well before the nuclear disaster. Accordingly, this chapter aims to unearth the roots of resilience in villagers' experiences of previous disasters and to illustrate how the experiential roots of resilience have been severed over time by state projects that have ignored the lived experience and workings of village life.[1]

Resilience is broadly defined here as the capacity of a community to respond to social and environmental shocks and distresses (Barrios 2016; Manyena 2006). This chapter draws explicitly on the life-environmentalist perspective, which brings three characteristic approaches to the issue of resilience (cf. Ueda and Torigoe 2012, Kanebishi 2016, Noda 2017). The first is a focus on small communities, such as villages and hamlets, as the basic geographic unit of analysis. The second characteristic is its interest in how resources from the commons are used in securing residents' lives and livelihoods in times of disaster (Furukawa 2006 [2004], Suga 2005). Thirdly, and most importantly, life-environmentalism adopts an "experientialist" approach. The experientialist approach asserts that understanding why individuals and communities act the way they do requires examining not the performed function of the act itself or actors' stated motivations but rather delving deep into the roots of the experiential world of the actors (Furukawa 2006 [2004]). Furukawa (2006, 33) states that experientialism "is at the core of life-environmentalism's breakthrough in modern knowledge."[2] In short, while existing resilience research often focuses on various contemporaneous conditions (e.g., demographics, physical environment, infrastructure, governmental services, economic well-being, and sociocultural capital) that enhance community resilience (e.g., Renschler et al. 2010), life-environmentalist research underscores the importance of the sociohistorical context in which such resilience has been cultivated and maintained, especially through the governance of resources on the commons, from the past to the present.

In the following, I first describe how the villagers of Kawauchi experienced a major famine over a century ago, and how they developed the forest commons to prepare for future disasters. I then detail how nineteenth-century modernization projects and the recent nuclear disaster and its aftermath have transformed and devastated these commons. The final section recaptures the highlights of the findings and draws implications for future policies and research on community resilience.

Nineteenth-Century Famine and the Experiential Roots of Resilience

To begin to understand the local history of disaster response, let us first look at a major disaster that struck the village 150 years ago through the eyes of a villager of that time. Girin Sakuma (1823–1899) was a leading farmer in the area who served as a *shoya*, or headman, in the late Edo period (1850s). He remains well known even today as a prominent leader in Kawauchi and as an intellectual who penned several books that recorded the events of his time for future generations. He is also regarded in his hometown today as a heroic figure for his efforts to help his fellow villagers and bolster communal survival when faced with the frequent and severe disasters of his time.

One of the most severe disasters of Sakuma's time was the Tenpo Famine of the 1830s, when several years of unseasonable weather led to poor harvests and widespread hunger. Historians suggest that this disaster killed roughly one million individuals, or around 3% of Japan's population at the time (Hayami 2001). Moreover, the northeastern Tōhoku region, where Kawauchi Village is located, was the area most seriously affected by this disaster.

Population data for the area clearly reveals just how seriously damaging the Tenpo Famine was for Kawauchi Village. At the village level, the population of Kawauchi was reduced to one-third of pre-famine totals. Looking at the hamlet level, we see for example that the population of Shimokawauchi, one hamlet within the larger Kawauchi Village, was 234 households and 1022 people in 1755, just before the famine. However, as shown by Yamaguchi (1974 [1940], 187), the number of households was greatly reduced after the catastrophic crop failure of 1836. Following the subsequent famine, registers record only 100 households and 300 people in the hamlet. As Yamaguchi points out, even in 1876 the population of Shimokawauchi had not recovered to pre-famine levels, with only 106 households and 815 people populating the hamlet.

Sakuma, who was still a teenager at the time, recorded the severe tensions that resulted from this period of famine and hunger.

> A 13-year-old girl living in Shimokawauchi village was unable to eat enough due to the famine. In need, she tampered with the precious food of her next door neighbor. However, just then, the owner of the house came home. He was furious, seized her and took her back to her house. He faced the girl's father and said, 'This girl is a nuisance who steals from the houses of others. Take care that she never steals again!' The girl's father came to the neighbor's house many times to apologize. But when he asked for reconciliation, the neighbor never accepted the apology.
>
> The girl's father was at a loss. One night, he got white rice from somewhere. He made a rice ball out of precious rice that he couldn't have afforded, and he offered it to his sullen daughter. The girl was pleased, but she just couldn't swallow the rice. That night, when she fell asleep, her father strangled her.
>
> Eventually, the father wrapped his daughter's body in a straw mat, slung the dead girl over his shoulder, carried her to the next house and screamed. 'I brought her like this to apologize for her stealing. Please forgive me!' He then threw his daughter's body into their kitchen. The family who witnessed this was stunned and felt shame for what they had done.
>
> (Sekisetsu Chihō Nōson Keizai Chōsasho 1935, 66)

After recording this event, Sakuma noted in summation that "Such events do not normally occur. These horrific events only happened because the villagers were thrown into extreme conditions by famine."

After spending much of his youth under such grim circumstances, Sakuma later began to try and understand why this catastrophe had occurred. What he found was that this disaster was not caused only by environmental factors such as unseasonable weather but also by social factors: namely, the profit-oriented agriculture pursued by farmers in the area.

> In the area where I live, there was a rice variety called *Jōkoku*. This rice is as an *okute*, or late-harvest variety of rice. Accordingly, there is always a risk that the harvest of *Jōkoku* will be diminished as a result of frost damage.[3] However, every house preferred to plant this variety because it yielded significantly more than other varieties. However, to our great sorrow, we faced the great famine of Tenpo as many farmers selected *Jōkoku*

and reaped poor harvests. Our area was severely damaged. The number of dead and out-migrants was shocking.

(Sakuma 1886, 105)

Historian Isao Kikuchi (2003, 386–7; 2019, 112–4), a leading scholar on the Tenpo Famine, has identified the relationship between famine and rice cultivation. He describes how feudal lords issued repeated bans on varieties of late-harvest rice to avoid frost damage and crop failure.

Despite these bans, Kikuchi (2019) states that the high potential yields of okute presented a risky, but potentially highly lucrative, proposition that both lords and farmers continued to find tempting and difficult to resist: "since rice was the most important commercial crop, it was produced through a tug of war between prioritizing measures against cold weather and prioritizing yield and profit" (Kikuchi 2019, 112, author's translation).

Amid this tenuous back and forth, Sakuma sought to produce crops that were suitable for the cold weather of the area, rather than high risk varieties. His zealous effort moved villagers, and more cold-weather resistant varieties were adopted in the village over the years. The communal experience of suffering from the famine had a decisive impact on Sakuma's consciousness and villagers responded to his call because it was understood that his effort was a selfless act. Sakuma's pursuit of forms of agriculture more suited to the natural environment did not end in the fields of the village, and expanded toward another site: the village's forested commons.

Cultivating the Commons, Nurturing Resilience

Sakuma's efforts were not limited to the planted fields of the village. He also sought to transform the mountainous forests of the area into a commons from which to gather and harvest "famine food" – food to nourish villagers in times of famines. Let us look at three ways in which Sakuma used these forested common areas to enhance community resilience.

Our first case shows how his experience of catastrophe led Sakuma to recognize the importance of famine food.

> In Kawauchi Village … when food is scarce, villagers go into the mountains and search for bracken. [During the famine] we made starch from its roots and prepared it for consumption, thereby managing to stave our hunger … from this starch we made an edible bracken mochi, or we saved the starch and used it as a preserved food. If there was extra, we would also put it up for sale.… Depending on the location, some bracken roots contain a high starch content while others do not. This distinction cannot be explained in words. I was 17 years old in the year of the Great Famine of Tenpo, and I dug bracken every day. That's how I supplemented my meager meals, and that experience penetrated my bones, leading me to have a good understanding of bracken root.
>
> (Sakuma 1892, 448)

Although Sakuma was born into a wealthy farming household, during the Tenpo Famine he found himself lacking regular meals. He had to roam the forested mountains in search of bracken root, an important famine food in the region. Bracken root digging, which should have been only an emergency measure, became painstaking routine work during this disaster. Through this experience of exploring the fields to survive the famine, Sakuma attained deep knowledge about famine foods.

A second example shows that Sakuma's experience of disaster helped him to recognize the need to prepare famine foods in preparation for disaster. That is, Sakuma sought ways not only to "gather" from the mountains in an emergency, but also to "cultivate" in preparation for a disaster.

> In the area where I live, when there is a bad harvest, we head into the mountains and collect the leaves of pokeweed [*Synurus pungens*]. These leaves are mixed with millet to make rice cakes, which help to survive hunger. However, at times of famine, it is difficult to find a single leaf because all the men and women in the village compete for what can be gathered in the mountains. That's why I keep the seeds in the fall every year and plant them in the wastelands, mountains, and other vacant lands. Even if you don't bother to go deep into the mountains, you can harvest about 100 kan [375 kg] every year…. It is good to mix them [after carefully processing the leave; author's note] with rice flour or other grains, and steam it to make rice cakes to eat. It is healthy and good for the stomach. There is no better food in the event of a bad harvest. People living in mountain villages should never make light of this famine food.
>
> (Sakuma 1888, 509)

In other words, Sakuma sought to "cultivate" famine foods to avoid competition for scarce resources in times of disaster. The amount of leaves harvested each year far exceeded how much the households of the harvesters themselves could eat in a year, thus creating a stockpile. It is difficult for all farmers, in particular poor farmers, to embark on such extra work from a long-term perspective. However, it can be seen that Sakuma, a high-ranking farmer, understood and attempted to improve the resilience of the local community by utilizing the commons to cultivate famine foods.

Our third case comes from records of the local response to this disaster and clearly indicates just how important the commons were to disaster response. In 1896, devastating floods occurred throughout Japan, and damages from these floods amounted to 70% of the national budget of the time. However, in Kawauchi Village, flood damages were mitigated by the community's deployment of resources from the commons:

> This year, Kawauchi Village also suffered a flood of about 3 meters and there were many victims from the floods throughout the village. However, damage to the fields was less than I expected. Some crops rotted due to

the long rains, but rice is likely to be harvested. In addition, chestnuts are ripening in the surrounding mountains. Every house had a large harvest of several *koku* [author's note: 1 *koku* is about 180 liters]. This year, the mountains were not open to other villages, and only commoners were given a certificate to gather chestnuts. We made a rule to collect small fees from those who actually picked up chestnuts. [Because of the establishment of a system in which the needy could live on the blessings of the mountains,] the villagers were happy despite the disaster.

> Sounds of chestnuts being weighed,
> Rustling in every house,
> Replacing our loneliness,
> Autumn evening.
>
> (Sakuma 1896, 439)

This description shows how the bounty gathered from the mountainous commons played an important role in suppressing the damaging effects of this disaster. While chestnut picking was allowed by anyone under normal conditions, during the disaster, voluntary rules were established to restrict gathering to local community members with the condition of paying a small fee.

What is important here is that these chestnuts did not just happen to be available. In preparation for such an emergency, the villagers systematically planted and protected these trees on the common lands. Figure 10.2 depicts the land use and land cover of Kawauchi Village in the eighteenth century, clearly indicating

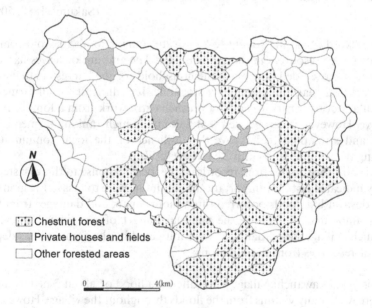

Figure 10.2 Locations of chestnut forests. Map by the author.

the abundant chestnut trees planted and maintained by villagers. In this way, villagers regularly and continuously managed nature – the commons – to secure resilience in the event of an emergency.

In life-environmentalism studies, this process, through which communities use the commons to secure their means of life, is known as the "right to life for the weak" [*jakusha seikatsu ken*] (e.g. Furukawa 2006 [2004]; Torigoe 1997). The right to life for vulnerable village members is a local rule that gives the needy preferential access to the commons, overriding existing social hierarchies within the village.

In other words, in addition to public relief systems and individual efforts, Japanese village communities have created and maintained locally specific disaster-response measures. Public relief systems and individual efforts alone were historically not enough to respond to disasters. That is why the unique disaster-response measures of specific local communities became so important, and the management of the commons has been a vital part of these measures.

Repeated Dispossession of the Commons

Since Japan embarked on modernization from the late nineteenth century, tight connections between local communities and the land have been strained and nearly severed through various forms of institutional reforms and developmental projects under the auspices of the centralized government. Let us look at two critical moments at which the dispossession of the commons was accelerated.

Dispossession Through the Nationalization of the Forest

The most damaging policy was the notorious nationalization of forest lands that began in the early Meiji era. Following the Meiji Restoration (1868), the new Meiji government sought to secure its financial base not only by transforming the lands of former feudal lords into national property but also by nationalizing common forests and grasslands beyond the confines of residential village spaces and agricultural fields (Kaneko 2017). While nationalization of private property would have drawn tremendous outrage, the state was able to strip customary (use and access) rights away and transform common forests and grasslands lands into state-owned land with the sweet-sounding and disingenuous promise of a reduced tax burden on those public lands.

Looking at this process more closely, forested land in Kawauchi Village was divided into four categories by the mid-Edo period. The first was land owned by the lord, or *ohayashi* ("the lord's forests"). Villagers were not permitted to log these forests, but only seven such forests were established in the village. The second category was *mochiyama* ("privately held mountains"), which were small forests close to agricultural fields and residences. These lands were privately owned or jointly owned by several households, and their resources were protected, planted, and felled as needed for daily life.

The third category was *kusakariba* ("mowing grounds"), areas relatively close to the residential spaces where villagers collected various types of grasses to be used as green fertilizers for fields, feed for cattle and horses, and thatch and other materials for residential construction and repairs. The fourth category of *tōyama* ("distant mountains") referred to all mountains and grasslands not included in the above three categories. All of the above forests outside of the lord's forests were frequently grouped together under the broader category of *sanno* ("scattered fields"), a richly suggestive term indicative of the existence of a vast area of lands held in forms of common use in the Village.

As mentioned, however, the Meiji government's efforts to redefine forest property rights resulted not only in *ohayashi* becoming state-owned land but also in the transfer of grasslands and forests that had been for common use such as *kusakariba* and *tōyama* into national property. As a result, of the vast forest area of roughly 17,000 ha in the village, only 260 ha were left under the control of the village community (Figure 10.3). Lacking access to forest resources, the lives of many residents of Kawauchi were cast into dire straits, and they filed proceedings against the national government for return of land to the village (Satō 1994, 357).

The subsequent court battle lasted from 1904 to 1911, with the village eventually winning their case against nationalization. However, of the approximately 15,000 ha that the village requested be returned to their control, only 6,500 ha were actually returned. Although only 45% of the area of their initial request, this was actually the second largest return area of the 1,538 such proceedings filed across Japan. These court battles were far from fair. Nearly 80% of the filed cases resulted in a verdict against the petitioners, without any return of land. In other words,

> [E]ven if it is possible to file proceedings, the hurdle for lands being returned was quite high and the case might be lost without gaining anything. It must be remembered that there are still quite a few towns and villages where state-owned land occupies more than 70% of the total area.
> (Satō 1994, 587)

A Partial Victory

Fortunately, Kawauchi won a historic victory against nationalization on November 21, 1911. This victory resulted in the reorganization of land ownership in Kawauchi Village into the following categories and areal composition: national forest (6,000 ha), village forest (10,000 ha), and private forest (1,000 ha).

A legal agreement based on modern law was created that stipulated how the lands returned to village control would be used. This agreement included provisions for how resources from the commons would be used in the event of a disaster. Namely, it stipulated that in the event of extensive damage caused by

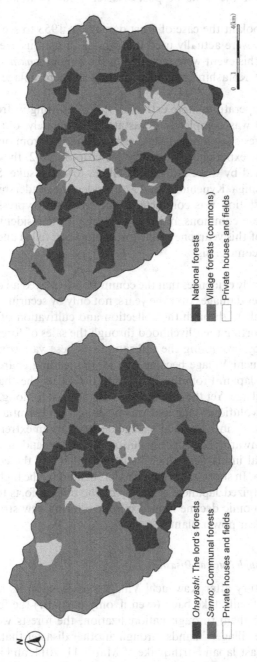

Figure 10.3 Changes in forest ownerships. Made by the author based on Yamaguchi (1938).

a disaster in the village, standing trees on these common lands could be sold and used to cover the costs of post-disaster recovery if agreed on by the community.

Let's take a look at the case of frost damage in 1953 to see how resources from these lands were actually used following a disastrous event. The procedures following this event were recorded in the annual *junkaichō* (community circular notes) of Kotashiro, a hamlet within Kawauchi Village.

> This year, temperatures were extremely cold, resulting in frost damage. The rice crop was completely devastated. Fortunately, our village has communal forests (commons) that we have inherited from our ancestors. Based on their experience of the terrible crop of 1902, these commons were established by three individuals, Shiga Matsunosuke, Shiga Kiyomatsu, and Shiga Kyuemon. In 1920, under the leadership of Shiga Kijiro, afforestation was conducted. Pine trees and cypress trees were planted in these commons and managed by all the residents. After the terrible crop of this year, one of the forest plots was felled and sold, helping us to overcome crisis.

This example clearly evidences that the commons has played an essential role in mitigating disaster damages over the years, not only by securing the survival of community members through the collection and cultivation of famine foods, but also by supporting their livelihood through the sales of forest products.

In addition, by promoting the utilization of these vast community forest resources, Kawauchi Village became one of the leading charcoal-producing villages in all of Japan. From the 1910s until the 1950s, the charcoal industry boomed in the village. Yet these boom days did not last for long. Following the postwar "fuel revolution," charcoal was no longer an essential resource, and the forestry industry in Kawauchi Village declined to an extreme extent from the mid-1950s onward. The fuel revolution had a tremendous impact on the charcoal and coal industries, which had been pillars of the economy of the Hamadōri region. In short, the region was left behind by the high-speed growth of the newly energized Japanese economy of the time. Efforts to catch up and reverse such economic decline hinged on one alluring new strategy: the construction of nuclear power plants.

Dispossession Through Nuclear Disaster

Exactly one century after Kawauchi Village succeeded in its legal victory in 1911 and could finally take back (even if only partially) the forests that had been seized from them through nationalization, the forests were once again wrested from the villagers' hands through another disaster that struck the village: the Great East Japan Earthquake of March 11, 2011, and its accompanying nuclear disaster. While radioactive materials spread widely through the area, fortunately the radiation levels in Kawauchi Village, which is only 20 to

30 kilometers away from the Fukushima Daiichi Nuclear Power Plant, remained at a relatively low level. The latest data from June 2020, based on the average of measurements from 4,040 separate points, reveal radiation levels of only 0.153 µsv/h [reference value 0.23 µsv/h]. In other words, Kawauchi Village meets the standards necessary to be deemed a habitable space.

However, what about the forests of the village? Even today, radiation levels remain at about 1 µsv/h (Taira et al. 2019). In addition, the decontamination of forests is not practical due to the enormous costs it would entail (Evrard et al. 2019). In other words, even now, a decade after the earthquake, there seem to be no effective measures to deal with forests in the village and villagers are left to simply await the passage of time (see Kaneko 2017).

Even in the absence of a nuclear disaster, timber prices had declined and the economic value of the village forests had been undermined well before 2011. It is also true that, after the disaster, TEPCO (Tokyo Electric Power Company) did compensate villagers for damages to contaminated forests. So, is it the case then that the issue of the contamination of the commons in village communities has been resolved? From a resilience perspective, the fact that the villagers have been disposed from the village commons for a second time is a critical and ongoing issue.

Lessons for Rebuilding Resilience: Looking Back to Move Forward

By pursuing a historical analysis of resilience in a specific community from the perspective of life-environmentalism, this chapter demonstrates that the commons has historically served as a safeguard for that community. Moreover, use of the commons stretched from a basic level of survival, as shown in the example of famine food gathering, to the level of sustenance, as shown in the examples of providing resources, and to funds for disaster recovery through planned use of these lands. Most importantly, this resilience was generated through the lived experience of local residents themselves, as they actively assessed their resources and confronted catastrophic conditions to find ways to best manage their environment to meet their needs.

Looking back at the history of Kawauchi Village's resilience through the years, there is no denying that the ability of the community to secure its own livelihood, cultivated through the use and management of the commons, has been seriously weakened. This chapter has clearly shown that the nationalization of the forests in the Meiji period and the nuclear disaster since 2011 have caused the local community to be severed from the commons twice over. It can be argued that both of these moments of dispossession occurred because of the national government's ignorance and indifference toward the local commons. In other words, for Japan's developmental state apparatus, forests were reduced to a vast and monolithic hinterland only visible within a market calculus. The result is that local mechanisms of resilience spontaneously developed by local communities have been devastated by policies radiating from the central government. In this way, a close scrutiny of the history of a local community from a life-environmentalist perspective can help reveal the functioning of power in place.

It has been frequently noted that – even ten years after the disaster – reconstruction from the Great East Japan Earthquake has been slow and halting. One reason may be that the roles played by the forest commons described in this chapter remain not well known. Those today who would uncritically accept the logic that forests of declining market value are outside the scope of decontamination measures would do well to recall the warnings of Sakuma from the Tenpo Famine. We must resist discourse and policies that cast contaminated forests as debts, and instead recall the experiences of our predecessors in order to reconceptualize the land as a still remaining asset.

Acknowledgment

This work was supported by JSPS KAKENHI Grant Number JP17KT0063 and JP17H02438.

Notes

1. The content of this chapter is based on Kaneko (2021) with substantial revisions. All block quotations in this chapter are translated from Japanese texts.
2. In some previous life-environmentalist studies, the word "empiricism" rather than "experientialism" is used as the translation of *keiken-ron* as it is the case in Furukawa (2006). However, in western philosophy, "empiricism" is usually associated with a natural scientific mode of knowing that values the role of experiments. Along with other chapters of this book, this chapter uses "experientialism" to distinguish life-environmentalism's approach.
3. Kawauchi, a mountain village, has a short harvest season with a frost risk.

References

Barrios, Roberto E. 2016. "Resilience: A Commentary from the Vantage Point of Anthropology." *Annals of Anthropological Practice* 30 (1): 28–38.
Evrard, Oliver, Patrick J. Laceby, and Atsushi Nakao. 2019. "Effectiveness of Landscape Decontamination Following the Fukushima Nuclear Accident." *SOIL* 5 (2): 333–50.
Furukawa, Akira. 2006. *Village Life in Modern Japan: An Environmental Perspective*. Melbourne: Trans Pacific Press. (Published originally in Japanese as Furukawa, Akira. 2004). *Mura no Seikatu Kankyōshi*, Tokyo: Sekai Shisō Sha.
Hayami, Akira. 2001. *Rekishi Jinkōgaku demita Nihon* [Japan seen from Historical Demography]. Tokyo: Bungeishunjū.
Kanebishi, Kiyoshi. 2016. "The Inner Shock Doctrine: Life Strategies for Resisting the Second Tsunami." *Institute of Social Theory and Dynamics* 1: 24–41.
Kaneko, Hiroyuki. 2021. "Nōgyō Zasshi nimiru Sakuma Girin no Nōgyōkan - Saigai o Ikinuku Nariwai no Mosaku to Teian" [Girin Sakuma's Perspectives on Agriculture Observed in *Nōgyō Zasshi*: Search and Proposal for Livelihoods to Survive Disasters]. *Journal of Agricultural History* 55: 97–109.
———. 2017. "Radioactive Contamination of Forest Commons: Impairment of Minor Subsistence Practices as an Overlooked Obstacle to Recovery in the Evacuated Areas." In *Unravelling the Fukushima Disaster*, edited by Mitsuo Yamakawa and Daisaku Yamamoto, 136–53. London: Routledge.

Kikuchi, Isao. 2003. *Kikin kara Yomu Kinsei Shakai* [Reading the Early Modern Society From Famines]. Tokyo: Azekura Shobō.

———. 2019. *Ue to Shoku no Nihonshi* [Japanese History of the Famine and Food]. Tokyo: Yoshikawa Kobunkan.

Manyena, Siambabala Bernard. 2006. "The Concept of Resilience Revisited." *Disasters* 30 (4): 433–50.

Noda, Takehito. 2017. "Why Do Local Residents Continue to Use Potentially Contaminated Stream Water After the Nuclear Accident? A Case Study of Kawauchi Village, Fukushima." In *Rebuilding Fukushima*, edited by Mitsuo Yamakawa and Daisaku Yamamoto, 53–68. London: Routledge.

Renschler, Chris S., Amy E. Frazier, Lucy A. Arendt, Glan-Paolo Cimellaro, Andrei M. Reinhorn, and Michael Bruneau. 2010. *A Framework for Defining and Measuring Resilience at the Community Scale: The PEOPLES Resilience Framework*, U.S. Department of Commerce, National Institute of Standards and Technology, Office of Applied Economics Engineering Laboratory, Gaithersburg, Maryland, Report NIST GCR 10-930.

Sakuma, Girin. 1886. "Okute o Waseshitsu ni Henzuru Keikenhō" [Methods to Convert Late Growing Rice to Early Growing Rice]. *Nōgyō Zasshi* 240: 105.

———. 1888. "Kyūkō Yobi Yamagobō Saibaihō" [Methods to Grow Pokeweed (*Synurus pungens*) as a Counter Measure to Famines]. *Nōgyō Zasshi* 319: 519.

———. 1892. "Warabi Nawa no Seihō" [Methods to Make a Rope from Bracken Ferns]. *Nōgyō Zasshi* 459: 448.

———. 1896. "Iwakino Kuni Naraha Gun Kawauchi Mura Tsūshin" [Report from Kawauchi Village in Naraha County, Iwaki Province]. *Nōgyō Zasshi* 582: 103–104.

Satō, Tokanori. 1994. *Kawauchimura Kokuyū Rinya Hikimodoshi Undōshi* [History of Recovering the National Forest in Kawauchi]. Kawauchi: Village of Kawauchi.

Sekisetsu Chihō Nōson Keizai, Chōsasho ed. 1935. *Tōhoku Chihō Kyōsaku ni Kansuru Shiteki Chōsa* [Historical Study of Famines in the Tōhoku Region]. Shinjō, Yamagata: Sekisetsu Chihō Nōson Keizai Chōsasho.

Suga, Yutaka. 2005. "Zaichi Shakai niokeru Shigen o Meguru Anzen Kanri – Kako kara Mirai e Mukete" [Safety Management Surrounding Resources in Local Societies: From the Past to the Future]. In *Kankyō-Anzen toiu Kachi wa* [Environment: What is the Value of Safety?], edited by Sumio Matsunaga, 69–100. Tokyo: Tōshindō.

Taira, Yasuyuki, Yūdai Inadomi, Shōta Hirajō, Yasuhiro Fukumoto, Makiko Orita, Yumiko Yamada., and Noboru Takamura. 2019. "Eight Years Post-Fukushima: Is Forest Decontamination Still Necessary?" *Journal of Radiation Research* 60 (5): 714–16.

Torigoe, H. 1997. "Who Gets the Most from the Commons." *Kankyō Shakaigaku Kenkyū* [Journal of Environmental Sociology] 3: 5–14. In Japanese.

Ueda, Kyoko, and Hiroyuki Torigoe. 2012. "Why Do Victims of the Tsunami Return to the Coast?" *International Journal of Japanese Sociology* 21 (1): 21–9.

Yamaguchi, Yaichirō. 1938. "Abukuma Sanchi niokeru Enkosagemodoshi no Kōyūrin ni Izonsuru Sanson no Keizaichiri – Fukushima Ken Futaba Gun Kawauchi Mura (Sono 1)" [Economic Geography of a Village in the Abukuma Mountains That Relies on Public Forests Returned from the State (Part 1)]. *Chigaku Zasshi* [Journal of Geography], 50 (5): 220–6.

———. 1974 [1940]. "Sanson niokeru Jinkō Chikan Genshō – Tōhoku Chihō no Sanson Kaihatsu to Jinkō Idō" [Population Replacement Phenomenon in Mountainous Villages: Development and Population Movement in Mountainous Villages in the Tōhoku Region]. In *Yamaguchi Yaichirō Senshū Dai 5 kan Seikatsu to Kikō Gekan*. Tokyo: Sekai Bunko.

11 Apparitions and the Recovery of Livelihoods after the 2011 Tōhoku Earthquake and Tsunami Disaster

Kiyoshi Kanebishi

> …the reconstruction of the town is very important. But, please do not forget that the lives of many people are still here and now. When people die, is that the end? I think about what those of us who survived can do.
>
> (Notes written on a desk found in a tsunami-hit area; Figure 11.1)

A large number of apparitions, or ghosts, have been witnessed in the areas of Japan affected by the 2011 Tōhoku earthquake and tsunami disaster. Such accounts are usually dismissed as curious or spooky stories, and are seen to have little to do with the post-disaster recovery of livelihoods in disaster-afflicted areas. Typical post-disaster recovery measures range from physical reconstruction of infrastructures to personalized psychological care (e.g., to cope with trauma), and to the creation of jobs and industries. Nevertheless, a question remains whether such measures alone are sufficient in securing the resilience of communities in the face of major disasters.

A key point of concern is the significance of "ambiguous loss" (Boss 1999). Pauline Boss, a family therapist, contrasts situations where the body of the deceased is available for a funeral and burial, called "clear-cut loss," to situations where someone remains missing with no clear point of death due to the likes of war, an aircraft accident, or a disaster, which she calls "ambiguous loss." The notion of ambiguous loss recognizes the distress and trauma experienced by bereaved family members whose loved ones are not confirmed to be living or deceased. This is a difficult situation for the bereaved because they cannot continue their lives "as they were," and at the same time they cannot restart new lives, either, because they have not completely accepted their loved one's death.

Under such circumstances, aggressive reconstruction measures that solely focus on those who survived the disaster, neglecting their connections with the deceased/missing ones, may leave them behind in the recovery process. This was precisely the situation after the 2011 disaster in Tōhoku, where many bereaved family members had to confront ambiguous loss.

At a glance, the apparition phenomena in the Tōhoku region may be understood as reflecting the trauma and internal struggles of the bereaved families

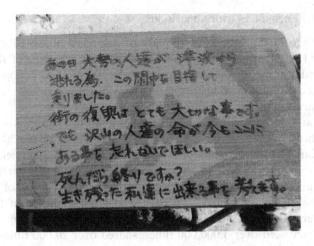

Figure 11.1 Notes written on a desk found in a tsunami-hit area. Photograph by the author.

that need to be resolved and cured. However, our fieldwork has shown that bereaved families and local community members often accept the apparition phenomena in a rather welcoming and affirmative manner. What accounts for their attitude that embraces the unscientific phenomena, and what may be the implications for post-disaster recovery?

Boss' work is suggestive in this regard as well. She argues that the manner in which people deal with, or their tolerance to, ambiguous loss is likely culturally formed, rather than being simple individual traits. She then provides an example of the Native American customs of treating those who die suddenly as still "living" for a period of time, in order to alleviate the sudden shock of their passing. This hints at the possibility of regionally and culturally specific ways in which the communities in the Tōhoku region deal with ambiguous loss.

In their study of disaster resilience, Zolli and Healy (2012) argue that resilience is about how communities *maintain their continuity* by accepting change and disruption, and by dynamically reconstituting themselves when faced with extreme environmental distress. Our study of a disaster-afflicted community in Tōhoku shows that the apparition phenomena and the community's response can be seen as a way to maintain the continuity of the community's life experience, connecting the past and the present, in the presence of ambiguous loss. In short, we came to understand that the long-term resilience of the livelihoods in the disaster-afflicted community is maintained by the ways in which the deceased continues to live with those who survived at least *for a while*. I argue that this understanding offers an opportunity to reflect on and shed new light on post-disaster reconstruction policies.

Taxi Drivers' Encounters with Apparitions

The magnitude 9.0 megathrust earthquake off the coast of the Tōhoku region on March 11, 2011, triggered massive tsunamis along the coast of the eastern Japan and nuclear meltdowns in Fukushima. A total of 18,423 people are confirmed dead or are still missing as of 2023; many of these deaths were the result of the tsunamis that swept a number of coastal communities.

The research project in disaster-afflicted communities in Tōhoku, which led to the publication of *Studying the Awakened Spirituality of the 2011 Tōhoku Disaster* (Kanebishi ed., 2016b), provoked much interest in such fields as sociology and theology, as well as among the general public. In particular, the newspaper article, which was first printed in the regional edition of the *Asahi Shimbun* newspaper on January 20, 2016, and later published in English as "Taxi Drivers Report 'Ghost Passengers' in Area Devastated by 2011 Tsunami," was accessed over 3,000,000 times, shared over 20,000 times on Facebook in just the first three days, and tweeted countless times as well. The story spread to as far afield as the USA, the UK, France, Russia, Brazil, and Taiwan.

Let us begin by brief examples of apparition phenomena from Ishinomaki City (pop. 150,000) in Miyagi Prefecture. Ishinomaki has flourished as a port town since the Edo period, and commercial fishing continues to be an important industry of the city. It is a sizable city, but it does have a sense of being a relatively closed-off, tightly knit community where everyone seems to know each other, and rumors spread quickly.

One of the interviewed taxi drivers ("SK," taxi driver, male, 56 years old) recalls:

> I think it was about three months after the disaster; I could check for sure in my records, but it was around the beginning of summer. Late one night, I was waiting for a customer near Ishinomaki Station, and a woman wearing a puffy coat like you wear in mid-winter got in my taxi.

The woman was in her 30s, and when the driver asked where she would like to go, she replied "to Minamihama." The driver got confused because the tsunami thoroughly destroyed the area.

Driver: That's now mostly vacant lots now; is that okay? Why Minamihama? Are you not too warm in your coat?
Woman: (In a trembling voice) Have I died?
Driver: Wha...?

When he looked in the rear-view mirror, no one was sitting in the rear seat.

On one afternoon in June 2014, another driver ("KH," taxi driver, male, 57 years old) was returning to the depot when he noticed someone hailing him. After stopping the taxi, a man wearing a paper mask got in. From his clothing and voice he seemed like a young man. The driver says, "But, the way he was dressed was—I mean—he was dressed for winter." When the driver asked him about the destination, he replied, "has my girlfriend been well?" The driver thought the customer might have been his acquaintance, and asked, "have we met somewhere?" The young man repeated, "has my girlfriend..." Before the driver realized it, there was no one in the backseat; instead, there was a small box with a bow attached. Since that day, the driver has kept the box in his car without ever opening it.

These taxi drivers had heard rumors of people seeing ghosts, but never thought it would happen to them. What surprised us was how these drivers eventually came to accept their experiences. For example, the former driver (SK) stated that

> But now I don't think it is anything particularly strange. Lots of people died in the Tōhoku earthquake and tsunami, right? It's natural that there are people who harbor despair and grief (*miren*). She [the passenger] must have been it [an apparition] ... I'm not afraid of them or anything. If someone wearing winter clothes out of season is waiting for a taxi again, I'll pick them up, and treat them as a normal customer.

Additionally, the latter driver (KH) recalled that

> I never thought it would actually happen to me. I was surprised, for sure. But still, nothing will change for me in the future, when someone raises their hand waiting for a taxi, I'll pick them up. Even if something similar were to happen again, I wouldn't stop and drop them off partway.

Other drivers whom we interviewed said that they would do the exactly same if they encountered a similar situation.

When they give rides to "ghost passengers," the fare will remain unpaid. The drivers foot the bill, and they will not report the incident to their employer because their boss will not believe the story anyway. Many of them do not even talk about those stories to their families or colleagues; they may have been initially frightened by the encounter with the apparitions, but appear to keep the experience to themselves as "good memories."

What interests us is not the scientific validity of the apparition phenomenon; rather, it is how people came to accept the phenomenon in a surprisingly positive manner. This observation invites us to explore how people in the disaster-afflicted areas have dealt with the living, the deceased, and something in between.

The Logic of Temporary Entrustment—Keeping the Ambiguous, Ambiguous

When people die, is that the end? Theologians would surely hold this question to be answered. It can be said that they cope with death not in a purely scientific way, but rather with answers prepared in a manner appropriate to their religion. The problem is that those answers do not necessarily resonate with the people of the disaster-affected areas.

One of the most profound questions of the bereaved of the 2011 disaster is: *where have our beloved family members gone?* Normally, confirmation of death is possible by meeting with the remains of their loved ones, but this is not possible in the case of missing persons. This is the state of "being on hold" where missing persons are not confirmed to be living or deceased, and Boss (1999) calls this an "ambiguous loss," which makes it difficult for the bereaved to "move on."

Such a state of "being on hold" entails two layers of crises. On the one hand, the bereaved family members do not know whether and when their beloved ones may someday come home (emotional crisis). On the other hand, in the Japanese cultural tradition, the spirit of the deceased can attain Buddhahood (*jōbutsu*) after the memorial service (*kuyō*), usually with their remains in a coffin. Therefore, missing persons cannot become truly "deceased" until the service is held (spiritual crisis).

In contemporary society, there is a tendency and desire to curtail the ambiguous space between the living and the dead as quickly as possible. Consequently, nowadays we often witness institutionalized religious rites such as collective funerals and memorial services after various disasters to demarcate the living (future) and the deceased (past), effectively sending the missing to the world of Buddhahood (*higan*: "that world"). In contrast, what we observed in the taxi drivers' encounters with apparitions and how they embraced them hints at a different consciousness at work. That is, it appears that rather than rushing to eliminate the realm of ambiguity, they are making the realm meaningful to them, and, this practice, we contend, has important implications for the recovery and reconstruction of the disaster-afflicted communities.

The practice of putting things "on hold" is part of our everyday lives. Philosopher Tatsuru Uchida, in his book *Death and the Body* (2004), uses the term "intermediate zones" (*chūkankō*) in describing how we process issues that cannot be resolved immediately. For example, when tidying up a computer's file storage, files that are no longer needed are sent to the recycle bin, and files that belong to certain topics and categories will be moved to appropriate folders. However, there are often files that cannot yet be sorted out, and they tend to be left on the desktop—a space of temporary entrustment—for the time being. We usually do this as a matter of course, and see nothing wrong with it.

When applied to the context of "ambiguous loss," how does the practice of using an intermediate zone contribute to the psychological care of the bereaved? Due to the magnitude of the disaster that abruptly took the lives of people away, there has been a great deal of anxiety, survivors' guilt, and self-doubt among those who survived, wondering, for example, what more they could have done in the time between the earthquake and when the tsunami hit. Despite the sustained mental struggles, state and external economic actors tend to gradually press afflicted communities to forget the deceased, move on, and complete reconstruction.

However, from the perspective of the survivors, it is not the government or business who should decide when to "clean up the desktop;" rather, they should have the autonomy to decide when to settle the matter and move on. We often assume that the faster the reconstruction is, the better. I contend, instead, that true reconstruction of survivors' lifeways is possible only when their autonomy is thoroughly respected even in terms of the timing and pace of reconstruction.

In this regard, the aforementioned narratives of the taxi drivers are worth revisiting. They say that they would pick up apparition-passengers again and again, should they encounter them. For those who lost their loved ones in the disaster, these words are reassuring because they imply that even if the drivers may not have lost their own friends or relatives, they would respect and support those who were lost to the disaster as members of the local community. These words imply that it is okay even if it takes ten, or even twenty years, for some people to feel ready to move on.

By allowing the ambiguous to remain ambiguous through temporary entrustment, the survivors are dealing with ambiguous loss by not fully dealing with it. Rather than seeing such a practice negatively as delaying the recovery, we should learn from it in an affirmative light (Kanebishi 2016a). Let us then explore the cultural basis underlying this method of coping with ambiguous loss.

Traditional Japanese Religious Perceptions of Apparitions

A good place to start is to examine how apparition phenomena in disaster-afflicted areas would be interpreted from religious theoretical perspectives. In the conventional view of Japanese Buddhism, religious rites such as funerals and memorial services are the means to clearly separate the deceased from our everyday world (*shigan*: "this world"), sending them to *higan* ("that world"), as indicated with bold arrows in Figure 11.2. Following this view, religious scholars such as Kōkan Sasaki (2012) conceive the apparitions witnessed in the disaster-affected areas as disembodied spirits (*reikon*), which appear in this world because the spirits have not attained the Buddhahood yet.

Sasaki (2012) consequently claims that the expected role of (Buddhist) religion is to help the deceased who are lost and stymied, tormented by the feelings of envy and bitterness, and are suffering because they have not attained Buddhahood to become settled and stable, be at peace, attain Buddhahood, and become ancestral spirits that watch over and protect their descendants. By so doing, he suggests, the apparitions in the disaster-afflicted Tōhoku regions will no longer haunt the survivors in the community.

I suggest, however, that our observations of the apparition phenomena do not neatly fit into this conventional Japanese religious thought. If the apparitions were the tormented souls of the deceased, those taxi drivers would have implored them not to appear again, or they would have pressed their hands together in prayer (*gasshō*) to exorcize the spirits or to send them to "that world." Instead, by suggesting that they would warmly welcome the apparitions if they appear again, the drivers have taken a receptive attitude toward meeting with them.

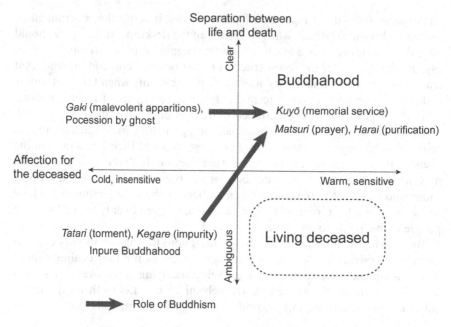

Figure 11.2 Schematic diagram of the positioning of the "living deceased." Created by the author based on Ikegami (2014) and other sources.

Our study therefore points to the realm of the "living deceased" (lower-right quadrant in Figure 11.2), in which temporary entrustment of ambiguous loss is permitted, if not welcomed. We argue that it offers an alternative view to the understanding of ambiguous loss, which is often seen as something to resolve and overcome (Boss 1999). In this way, our study also departs from the traditional Japanese religious perspective.

The Consciousness Behind and Cultural Foundation of "Living with the Deceased"

Let us further articulate the nature of the interactions between the living and the deceased in the present study's context. How do the bereaved family members continue to "live with" their beloved ones, whose bodies no longer exist? In what state of consciousness are they?

Critic Eisuke Wakamatsu (2012a) argues that the sense of sorrow that (living) people feel when faced with someone's death is not about the emotional conditions of the living; rather, it is a sign that the deceased is touching the soul of the living. He faults modernity for treating the sense of sorrow as pitiable, inconsolable, and unredeemable. He instead argues that sorrowfulness is the proof that the deceased has had a profound impact on one's life. We cannot see the deceased, but we sense their presence even more strongly through the feeling of sorrowfulness. Wakamatsu (2012b) goes on to say that "being with the deceased" does not mean to live everyday remembering the deceased; rather, it means to live with the deceased in the present moment. Wakamatsu call such a mode of being as being "invisible co-living neighbors."

For example, a mother whom we interviewed, who had lost her son to the disaster (but his body was still missing), pointed to a box of sweets that she had laid in front of his grave, and said that the box fell three times even though there was no wind. She continued, "he was always mischievous, and I am sure he came to play tricks on me again," and "I am much more scared of the living; I don't feel scared a bit if the dead were to appear." This example illustrates that in the mother's consciousness, the deceased son was still with her, which is consistent to Wakmatsu's assertion about the dead and the living being "invisible co-living neighbors." It is also noteworthy that the personality of the deceased person is maintained from his former life. This characteristic contrasts with a typical apparition phenomenon, which either shows no personality per se, or only comes with a sense of a violent grudge. For the bereaved people, the apparitions that they encountered in the post-disaster Tōhoku are better conceived as the "living deceased" who may be invisible for the most part, but who are as real as living people with their distinct personalities.

The stories of the taxi drivers and the mother show that they are dealing with apparitions as "living deceased," treating them as if they are living persons. Do they do so based solely on their individual past experience? I point out that the practice of keeping around the spirits of the deceased is linked to the historically accumulated experience in the communities of the Tōhoku region. In other words, there is a regional and cultural foundation underlying the practice.

Religious scholar Hiroo Satō (2015) discusses the practice of *mukasari ema* found in parts of the Tōhoku region. This is a particular type of *ema* ("picture horse"), or a votive horse tablet displayed at Buddhist temples, which depicts a child who died at a young age. On the wooden tablet, the child is drawn as a bride or groom at a wedding ceremony, one of the most important life events, that never actually happened. It is implied that the child, who tragically died before their parents, will continue to live their lives with their lifelong companion in another world. The Tōhoku region also has the distinct tradition of *kuchiyose*, a type of trance mediumship, in which the deceased is called down to this world through the medium and speaks to the bereaved family. In the aftermath of the 2011 disaster, *kuchiyose* ceremonies were seen in various parts of the region.

These practices indicate that there is already a cultural foundation in Tōhoku to embrace apparitions as part of their everyday lives. This may be contributing to the observed consciousness and behaviors of the residents of the Ishinomaki area, such as those of the taxi drivers who say, "[the community was] hit so hard by the disaster that it would not be so strange for apparitions to appear, and I would totally understand their feelings if they were to appear" (Taxi driver SK). In the relatively close, tight-knit community of Ishinomaki, there appears to be the feeling of sympathy and care among the residents, even if they may not know each other directly, and it is cultivated through the shared experience of losing their loved ones to the disaster.

Conclusions

Through the example of the apparition phenomena in post-disaster Tōhoku, this chapter has explored the meaning of the deceased continuing to exist in the present despite having already been publicly mourned into the past. Rather than seeking scientific explanations of apparitions or treating them as discrete individual phenomena, my goal has been to articulate the historically formed consciousness in the Ishinomaki community that allowed residents to embrace these apparitions in this particular way.

It has become apparent that the apparition phenomena must be understood as coping methods of survivors over "ambiguous loss" in the way that is grounded in their lifeways in Tōhoku. Somewhat analogous to what Boss (1999) writes about the Native American customs of treating the deceased as "living" for some time, I have elucidated how the communities of the Tōhoku region of Japan have their own ways of preserving the personalities of the "living deceased," which helps to absorb and mitigate the shocks of and distresses arising from calamities.

These coping methods are intricately linked to the long-term resilience of disaster-afflicted communities. There is currently a tendency to promote rapid reconstruction at all costs, which also means to send the souls of the deceased into nirvana as quickly as possible. This study has shown, in contrast, that keeping the spirits of the deceased around "for a while" is as an important life-strategy for bereaved families that allows them to postpone what cannot be resolved immediately and to diffuse the shock over a longer period of time (until they are finally ready to move on). This understanding provides a foundation for better post-disaster recovery policies that the survivors can fully embrace and their agency are respected, rather than resorting to the currently dominant "rebuild-and-move-on" style of recovery.

At present there is a tendency to affirm "rapid reconstruction" in which the souls of the deceased are swiftly sent into nirvana. Rather than such a rebuild-and-move-on style of recovery, this study has shown the possibility of a recovery that the survivors can take time and fully embrace. To that end, keeping the spirits of the deceased around "for a while" can be seen as a sensible life-tactic for the survivors to cope with the disaster. Such possibilities and life-tactics prompt us to envision more gradual reconstruction policies that take the emotional well-being of local communities into consideration.

References

Boss, Pauline. 1999. *Ambiguous Loss: Learning to Live with Unresolved Grief*. Cambridge, MA: Harvard University Press.

Ikegami, Yoshimasa. 2014. "Segaki (Buddhist Service for the Benefit of Suffering Ghosts) as a Subject of Religious Studies", *Komazawa Daigaku Bunka*, 32, 69–94. In Japanese.

Ishibashi, Hiedeaki. 2016. "Taxi Drivers Report 'Ghost Passengers' in Area Devastated by 2011 Tsunami," *Asahi Weekly*, April 3, p. 17.

Kanebishi, Kiyoshi. 2016a. *Shinsaigaku Nyūmon* [Introduction to Disaster Studies]. Tokyo: Chikuma Shinsho.

Kanebishi, Kiyoshi. ed., 2016b. *Yobisamaseru Reisei-no Shinsaigaku* [Studying the Awakened Spirituality of the 2011 Tōhoku Disaster]. Tokyo: Shinyō Sha.

Sasaki, Kōkan. 2012. *Seikatsubukkyō no Minzokushi – Dare ga Shisha o Shizume, Seija o Anshin Saserunoka* [Ethnography of Buddhist Lifestyles - Who Will Appease the Deceased and Bring Peace of Mind to the Living?] Tokyo: Shūnjusha Publishing Company.

Satō, Hiroo. 2015. *Shisha no Hanayome – Sōsō to Tsuioku no Rettōshi* [The Bride of the Deceased – A History of Funerals and Memory Throughout Japan]. Tokyo: Genki Shobō.

Uchida, Tatsuru. 2004. *Shi to Shintai – Komyunikēshon no Jiba* [Death and the Body – The Magnetic Field of Communication]. Tokyo: Igaku Shoin.

Wakamatsu, Eisuke. 2012a. *Tamashī ni Fureru – Daishinsai tto Ikiteiru Shisha* [Touching Spirits - the 2011 Tōhoku Disaster and the Living Deceased]. Tokyo: Transview.

———. 2012b. *Shisha tono Taiwa* [Conversations with the Deceased]. Tokyo: Transview.

Zolli, Andrew, and Ann Marie Healy. 2012. *Resilience: Why Things Bounce Back*. New York: Simon and Schuster.

Part IV
Historic Environment and Urban Communities

12 Living Traditional Culture
Gujo Dance in Hachiman Town, Gujo City, Gifu Prefecture, Japan

Shigekazu Adachi

In densely populated parts of Asia, such as Japan, humans have systematically modified nature to sustain their livelihoods. Accordingly, when Japanese social scientists, particularly environmental sociologies, address the issue of "protecting nature" in the contemporary context, the object has to be the nature that has been already "worked out" by humans. Life-environmentalism, in particular, most strongly emphasizes the historically accumulated practices of careful human interventions to the nature that has protected both human livelihoods and their environment at the same time.

Life-environmentalism positively values some degree of human modification of nature. Rather than putting nature and culture in a binary opposition, they are placed on a continuous spectrum. Therefore, the subject of life-environmentalist research is not limited to the preservation and sustainability of quintessential "natural" environment (e.g., lakes and forests), but extends to those of archaeological sites, historical townscapes, and cultural heritages. Life-environmentalism calls the latter "historic environment" and regards it as its import research object (Torigoe 1997). Traditional cultural practices such as festivals and folk performing arts, the main focus of this chapter, constitute intangible, but essential, aspects of such historic environments.[1]

Questions About Traditional Culture and Tourism

In the 2000s, the Japanese government declared a policy of *Kankō Rikkoku* ("tourism-oriented nation") and positioned tourism as a key driver of the national economy. At that time, festivals and folk performing arts in various regions of the country were highlighted as potential tourism resources, supposedly symbolizing the historically rich, "traditional cultures of Japan," that would attract foreign tourists. Within Japan, already from the latter half of the 1980s, local economic development through tourism was touted, and local traditional cultures were increasingly turned into tourism resources. As a result, Japan's local traditional cultures have been increasingly "touristified" for both domestic and international visitors.

Rather than lamenting these traditional cultures becoming increasingly commodified for tourist consumption, cultural constructionist research, particularly Eric Hobsbawm's work on the "invention of tradition," has revealed that "traditional cultures" such as festivals and folk performing arts had never been "authentic" in the sense of being inherited from the past unchanged, but have been invented or reinvented in the modern and contemporary political economic contexts (Hobsbawm and Ranger, 1983; Bruner, 2005). Nevertheless, rather than rejecting such invented traditions, cultural constructionists rather positively evaluate the process whereby local residents establish their own unique identity through their interactions with tourists who come to experience the "invented traditional culture."

Gujo Odori ("Gujo dance") in the Hachimancho area of Gujo City, Gifu Prefecture (hereinafter referred to as the "Gujo Hachiman" area), is also a typical example of such traditional culture. Gujo Odori is a general term that consists of ten dances, such as "Kawasaki" and "Haru Koma," and is originated from the dance during the *Bon* Festival (the annual summer memorial service for ancestors, practice throughout Japan) by local residents. However, due to the Meiji government's modernization/westernization policy, the Bon dance became prohibited for some time, supposedly because it would corrupt public morals. Eventually, the dance was revived as Gujo Odori and has been repeatedly "reconstructed" by the efforts of local residents, and now attracts about 300,000 tourists to the area each summer. Moreover, since it was designated as a National Important Intangible Folk Cultural Property in 1996, this dance has become the basis of local identity as well as a valuable tourism resource. Based on these observations, Gujo Odori may look like a model case of "invented traditional culture" as discussed by Hobsbawm and Ranger (1983), and also a desirable form of regional traditional cultural preservation through tourism.

However, during my repeated field visits, I began to hear some residents talking about "locals losing interest in Gujo Odori," and noticed that locals were watching jubilantly dancing tourists in distance at the dance venue.[2] Furthermore, some locals had invented a pre-touristified form of dance, which differs from the current Gujo Odori, and are even holding organized events around the "old dance." From an outsider's point of view, it may seem perplexing why they have taken such a series of actions, even though Gujo Hachiman already has Gujo Odori that the locals should be proud of. It is this question that leads us to explore and learn from the everyday life of the local residents, rather than resorting to the analytical viewpoint of "invented traditional culture."

These observations and local initiatives are indicative of an alternative way of preserving traditional culture that does not depend on tourism promotion. In this chapter, by analyzing the declining local participation in Gujo Odori and emergent dance events that differ from Gujo Odori in the Gujo Hachiman area, I draw practical implications for the creative preservation of traditional culture and historic environment.

Historical Changes of Gujo Odori

An Overview of Gujo Odori

Gujo Hachiman is a small town with a population of about 13,000, located in the inland area of Gifu Prefecture in the middle of Honshu, about 100 kilometers upstream from the mouth of the Nagara River, one of the three main rivers that flow through the Chubu region. It is an hour-long drive to the north from Gifu City, the prefectural capital. Gujo Odori, which is said to have a history of about 400 years, has been held in this town where wooden houses are densely packed in a basin surrounded by mountains on all four sides.

Gujo Odori takes place every year from mid-July to early September at night for about 30 days (of which four days during the *Bon* season are all-night dancing sessions). Each *jichikai* (neighborhood association) in the town is responsible for organizing events on a rotating basis during the 30 days. Each day goes like as follows. After conducting their own district events, the *jichikai* in charge of the day's dance sets up a large mobile dance house, called *yakata*, on the main street in their district. A *yakata* is like a two-story float on which a leading singer and players of the *shamisen*, flute, and drums perform (Figure 12.1). Before the performance starts at 8 pm, members of the local volunteer

Figure 12.1 A *yakata* float at the Gujo Odori festival. Photograph by the author.

group, Gujo Odori Preservation Society (hereinafter referred to as the Preservation Society), who are the main bearers of the Gujo Odori events, arrive at the venue. At the request of each *jichikai*, they play a musical accompaniment on the *yakata*, wearing matching *yukata* (informal cotton kimono), and dance around the *yakata* to show visitors a model dance. When the time comes, many visitors and local residents, along with the members of the Preservation Society, dance and move around the floats, forming big circles. The dance lasts for about three hours after which the *jichikai* members clean up the site and the night's event ends.

Bon Festival Dance as the Origin of Gujo Odori

To understand how the Bon Festival dance, originally a cultural practice for the local residents themselves, was transformed into such a popular event, Gujo Odori, attracting thousands of tourists each year, we need to look back the history of the region. In Gujo Hachiman, a castle was built in the Middle Ages, and many samurai, craftsmen, and merchants settled in the castle town since then. Especially in the Edo period (1603–1867), enormously wealthy merchants began to appear in the town through river trade. They traded goods and supplies by connecting villages in the region and cities in the downstream of the Nagara River, which runs through the western part of the town. These wealthy merchants were locally called *machishū* ("townspeople"), and they climbed up to the political and economic leaders of the town by donating money to the town, and even by lending money to the Gujo Domain (*han*) in the late Edo period.

During the Meiji period (1868–1912), Gujo Hachiman was the political and economic center of the Gujo region. Modern government offices such as courts, police stations, and post offices were concentrated here. Economically, raw silk, spun in large quantity in nearby villages, was an important means of earning foreign currency to modernize the country, and brought much wealth to the town. The wealthy, townspeople of Gujo Hachiman poured the enormous profits from the silk trade not only into their family business, but also into their pastimes, which ranged from *haiku*, *renku*, literature, tea ceremony, flower arrangement, painting, plays, photography, and folk art collection. They not only became connected to the literary and art scenes in Tokyo, but some also even engaged in cultural production themselves. As a result, Gujo Hachiman, even though it was a small inland town, became a highly sophisticated cultural center, comparable to Nagoya and Gifu Cities.

The economy of Gujo Hachiman began to fade in the Taisho period (1912–1926), however, because of the decline in the silk industry and river trade. Heavy industry took over such light industry as silk production, becoming the leading sector of the national economy. The town's river trade also declined due to the development of the railway networks, leaving Gujo Hachiman, which was without railways at that time, completely behind of and isolated from major cities. Even worse, a major fire burned down a large extent of Gujo Hachiman in 1919; this was a major blow to the town.

A way out from such a plight was found to be the touristification of the entire town. From the Taisho period to the early Showa period (1926–1989), Gujo Hachiman reinvented itself as a tourist town promoting such attractions as cherry blossoms, the castle and beautiful streams, and the biggest of all, the Bon Festival dance, which was reinvented as Gujo Odori.

Three Changes Leading to "Waning Local Interests in Dancing"

In 1923 the *machishū* townspeople and selected other locals established the Preservation Society for the revival of Bon dance, as the official prohibition of Bon dance by Gifu Prefecture (and the State) since 1874 became loosened. In the process of the revival, three key changes were made to the traditional Bon dance: sanctification, touristification, and formalization of training. These changes, however, had unintended consequences—namely the loss of interest among the locals in Gujo Odori.

First, in order to more readily get the state power's approval for the revival of the folk performing arts, the Preservation Society removed obscene lyrics from the dance songs, and reinvented Bon dance as a "healthy entertainment." For example, the minutes of the Preservation Society in 1925 read:

> To the dancer:
> The movement is sacred
> The dance is a manifestation of pure humanity
> It is also an expression of male and female harmony
> - We shall not sing obscene songs
> - We shall not corrupt public morals
>   (Gujo Dance Preservation Society 1922–1946, 13)

The Preservation Society apparently worked hard to enlighten the local residents of the sacred nature of the local performing art. The prescribed sacredness was inevitable to avoid conflicts with the police and state power, but as a result, the lyrics of the dance became fixed, and the improvised nature of lyrics and dance was gradually lost.

Second, the sanctification of Bon dance, as Gujo Odori, not only helped to gain an approval from the state power, but it also promoted the tourisification of Bon dance. Under the catchphrase of "healthy entertainment," the "orthodox" version of the dance, created by the Preservation Society, was already recognized among the locals as a potential tourism resource even before WWII. Nevertheless, it was during the post-WWII folk music boom, which amplified the exposure of local folk dances in media and various contests, that the number of tourists seeking to experience Gujo Odori rapidly increased. In response, in 1953, the town government and the Preservation Society made a decision to change the form of Gujo Odori in order to accommodate the growing number of tourists. Until then, the locals formed multiple circles of a small number of people and danced while singing (locally called *mukashi odori*, or "old dance"). This was changed to the current form, where a mobile *yakata* float with a singer and instruments is placed in one large circle of dancers, including visitors. In this change, the Preservation Society took on the active role of controlling the dance circles. Specifically, if the local people were dancing in small circles on their own as they had used to do, the Preservation Society staff would come and tell them to join in a large, "official" dance circle. This caused occasional skirmishes between the Preservation Society staff and the locals. Despite these issues, the Preservation Society forged ahead and promoted the standardized dance form, through such means as creating a manual book for how to dance properly, in order to meet the needs to tourists.

The third change was the formalization of musical training. The introduction of the *yakata* float, accompanying the touristification of Bon dance, further separated the musical accompaniment and the dance, and gradually robbed "voice" from the locals who did not belong to the Preservation Society. How did that happen? In the pre-WWII time, virtually any local residents could both dance and sing, and those with a high-pitched, beautiful voice could take the lead. During the dance, the lead singer is who moves the dance circle. That means, all the locals were eligible potentially to take the lead in a dance circle, implying that they could all be an agent of handing down the tradition. However, the introduction of the *yakata* float after the war led to the monopolization of lead singers to the members of the Preservation Society. One may say that those who want to sing should just join the society, but it was not so easy because the Preservation Society adopted the apprenticeship system to train successors of lead singers. With a strong pride of "we are who lead the dance," the teaching of the masters of singing and musical accompanists was extremely strict. Despite being an amateur performing art (as the world "folk art" implies), this professional-level, strict guidance acts as a barrier for most residents to join the Preservation Society. At the same time, this formalized training system creates factional disputes, for example, among the accompanists over the interpretation of the teachings of the masters, or between the accompanists and the lead dancers over the initiative during the dance. Hearing these factional disputes, local residents stay away even more from the Preservation Society.

These three simultaneous changes in Bon dance, sanctification, touristification, and formalization of training, have certainly made it possible to preserve and sustain Gujo Odori. Especially in the postwar period, the Preservation Society maintained its stature in the town by attracting tourists to Gujo Odori. The "monopolization" of Gujo Odori by the Preservation Society was most directly for tourists, but there was an underlying assumption that it would also benefit the local residents by extension.

It became evident, however, that the local residents who did not belong to the Preservation Society were deprived of their voice during the dance, and they were made to dance in the style that onetime visitors could acquire immediately. In the crowded circles pushed by tourists who moved their bodies "like robots" without singing, locals no longer found place to dance freely and enjoyably. This is how local residents have moved away from the dance that was supposedly their own.

Nostalgia and the Invention of a New Culture

Aesthetic Reality of Fuzei (Natural Elegance)

The locals speak about "waning local interest in dance," and lament that they cannot find their place in the current touristified Gujo Odori. In turn, they favorably talk about the "feel" of the dance venue before it became a tourist attraction, especially before WWII, saying "those were the good old days" and "I miss the old days." Why such nostalgic remarks? In the old days, one could always find their local friends or acquaintances dancing in a circle at the venue and they ask you, "let's dance together." Then, the friends will effortlessly bring you into the dance circle. An older woman also remembers that when she was young, on her way to dancing with her friends, they met a group of male friends. The boys asked, "Where are you going?" And the girls answered, "Of course to the dance!" Then, they all say, "let's go together," and join in the dance circle. One of the male friends took charge, saying "you all line up nicely like, boy, girl, boy, girl... ok?" and we danced all together. She recalls that the naturally ordered ambiance felt really nice.

Let us also look at how "old" dancing among local acquaintances differs from the current form of dance in terms of the dynamics of the dance circle itself. In short, the former embodies two contrasting characteristics: competition and integration. As mentioned in the previous section, before the touristification of Bon dance, everyone could sing and dance, so that during the dance they would watch vigilantly for a chance to take the lead singer position and show off their singing skills. They would not care about detailed hand movements, perhaps dancing with just with their feet, and eagerly aiming for the lead role. At the moment when the previous singing is finished, multiple singers start singing to take the lead like a musical chairs game. A person with a high-pitched, loud, long-lasting voice who starts the first line of the song can then become the next lead singer. The tense atmosphere of not knowing who will take the next lead makes the dance even more engaging.

In the "old odori" it is also required for a lead singer to improvise lyrics as they sing. Locals gather around a circle in which the lead singer improvises interesting lyrics and sings with a beautiful voice (while still dancing as well). As the circle grows larger, the lead singer stops dancing and moves into the circle. The singer then continues to sing while walking in the opposite direction from the movement of the dancers. However, as one singer continues to dominate the big circle, their voice may no longer reach the entire circle and the dance may begin to be out of sync. At that time, some dancers may say, still singing along with the song, "This dance is no longer fun!" and break out of the circle with a few friends, creating a new circle to dance to another song. In this way, the "old odori" venue was quite fluid and dynamic in form, being influenced by who took the lead, how many circles were formed, and how big the circles became. Locals enjoyed the competition over who sings well within a dance circle, or between multiple circles.

While competition was an important aspect of the "old dance," there was another aspect, integration, where the entire dance eventually became a "one whole." This symbolically appears in the scene of the "all-night dance" during the four-day Bon period. A local resident living near the all-night dance venue recalls:

> During the all-night dance, by the time the sky is getting lighter after dawn, tourists are no longer around as they returned to the inn, and only local dance enthusiasts are still dancing. That dance is beautifully in sync, maybe like swans. And there, the first bus of the day passes by the dance circle.

Even the locals who had little interest in dancing would get up early and check out that scene—it was such a special moment shared by all residents of the town.

Local residents fondly talk about the special feeling that arose from the facts that they enjoyed dancing freely and orderly at the same time among acquaintances; singers thrillingly competed for their skills; and the dance became beautifully synthesized and synchronized in the end. Locals still talk in their daily conversations about how "there was natural elegance" (*fuzei ga atta*) in the old odori. Importantly, they are not lamenting the current state of Gujo Odori; rather, by saying, "those were the old days" and "I miss those times," they are constructing an intersubjective reality based on the aesthetic past, which could be summed up as *fuzei* (風情: translated here as "natural elegance").

From the Past as an Ideal to the Creation of a New Culture

The reality of "natural elegance" constructed through nostalgic narratives by local elderly people is certainly interesting in itself, but some cultural theorists may consider such a retrospective view, which praises the past that never returns, as ultimately unproductive.

Folklore scholar Shōji Inokuchi's discussion of the Japanese word *mukashi* ("the past") is useful when thinking about the significance of the common daily practice of feeling nostalgic about the past. According to Inokuchi, the word *mukashi* has two meanings. One is "the past as a historical concept" (Inokuchi 1977, 6), and the other is "the past as an ideal" (Inokuchi 1977, 9). The former is a common-sense notion; it is a point on a linear temporal axis; it is "the past" that one can never return to. The latter meaning—"the past as an ideal"—embodies the normative view that idealizes and affirms the past, which leads to the belief that bringing back the past is the best way to reform the society (Inokuchi 1977, 10). What is noteworthy in this discussion is that this notion does not simply advocate for "going back to the past;" rather, it has the nuance of "something that *may* have existed in the past, but that should definitely be created in the future," and "a goal that we should strive for…and it is an imagined and dreamed world" (Inokuchi 1977, 11).

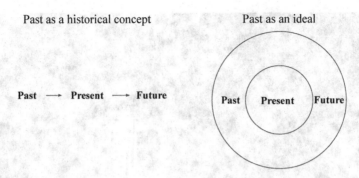

Figure 12.2 Two concepts of "past." Created by the author based on Inokuchi (1977, 11)

If we follow Inokuchi's discussion, the local residents' conversation about the dance comparing the past and the present does not only make the conversation participants critical of the present, but it also gives rise to the potential to mobilize themselves, while feeling nostalgic, to actively construct the ideal form—what the dance should be—toward the future. This logic is captured in Inokuchi's diagram of the two meanings of "the past." In the conception of the past as a historical concept, "the past" is divided by "the present" from "the future." On the other hand, in the conception of the past as an ideal, "the past" is meshed with "the future," surrounding "the present" in the center (Inokuchi 1977, 11–12) (Figure 12.2). As an extension of this logic, "living traditional culture" can be formally thought of as a present practice enfolded by an ideal past, which is at the same time a desirable future.

It is indeed evident that some of the Gujo Hachiman locals are driven by the aesthetic reality of "natural elegance." In September 1996, an event called "Old Dance Evening" was held for the first time at a local temple between the official Gujo Odori events. The organizers of the event wanted to recreate the "naturally elegant" prewar dance by "going back to the basics."

Their dance did not use any *yakata*, musical accompaniment, electric lighting, or microphone, but uses only one "Kiriko lantern" (the spirits of ancestors supposedly gather here) hanging from above (Figure 12.3). At the start of the event, after all the participants prayed toward the lantern for the ancestors' spirits by the organizer's signal, the participants danced the "old dance" with multiple small circles while singing on their own.

At the dance event, there was supposed to be competition for lead singers as described above; however, the participants could not handle the successive turnover of lead singers well and the dance was occasionally interrupted. Looking at such a situation, the organizers painfully realized how they had been deprived of their voices and songs because they had all been accustomed to Gujo Odori that separated dancers and accompaniments for many years. Consequently, they decided to form the "Gujo Odori Ohayashi Club," independently from the

Figure 12.3 Kiriko lantern. Photograph by the author.

Preservation Society, and began to work on their own singing and accompaniment lessons. Members of the newly formed club teach each other skills and techniques, undoing the formalized training system based on the master–disciple relationship. Since its formation, as the skill levels of the Oayashi Club members advance, they have been regularly invited to perform in events in and out of Gujo Hachiman.

Especially in Gujo Hachiman, various dance events are now held for "locals wishing to dance comfortably" and "bringing back the easy-going dance" on days other than the officially scheduled Gujo Odori events by the Preservation Society. In those events, the Oayashi Club members often volunteer to provide musical accompaniment, and the club has become the key actor in "non-official" Gujo Odori events.

These emergent activities such as "Old Dance Evening" and "Ohayashi Club" are appealing to all local residents to regain their interest in the dance with the slogan "Take back the dance in our hands," which is rooted in the locally shared aesthetic reality of "natural elegance." Furthermore, these cultural activities are indicative of the locals' self-reflective attitude in reaction to the ceaseless touristification of their dance. At the same time, they show the way forward as to how they wish to inherit their traditional culture that puts their enjoyment in the center. Accordingly, the nostalgia seen in the daily conversations by the locals is not a sign of backwardness; rather, it is fundamentally future-oriented and facilitates the creation of a new culture.

Sustaining Traditional Culture Centered on Local Enjoyment

In this chapter, I have explored a potential pathway of sustaining cultural tradition by drawing on the local experience of the Gujo Hamicham community, in which the locals had apparently lost interest in Gujo Odori, and began to hold their own, alternative dance events. That pathway centers on the enjoyment of the local residents themselves, rather than being driven by tourism, as seen in many regions of the country. In short, my study shows that building and sustaining a better culture cannot be divorced from locals' enjoyment. Regarding this point, the words of Kunio Yanagita, the founder of Japanese folklore studies, are illuminating:

> I happen to believe that a joyful life is precisely what culture should be all about, and that enhancing the joy even further is what improves culture.
> (Yanagita [1941] 1970, 201; translated by the author)

Here Yanagita openly states that "culture" is synonymous with "joyful life." In the same way, this chapter has shown that "living traditional culture" cannot be separated from the enjoyment of the Hachiman Gujo residents, who after all are the most likely bearers of the culture.

In addition, in the context of the present study, what led to the "enhancement of culture" was the aesthetic reality of "natural elegance," which emerged from retrospective conversations among local residents feeling nostalgic about the past. That aesthetic reality acted like a blueprint, to which some of the locals responded and invented such new culture as the "old odori," overcoming the problems of the current cultural form. From a life-environmentalist perspective, the nostalgic narratives of the locals should not be overlooked merely as a "backward-looking" sentiment; rather, they can be seen as an opening for change and active local agency.

Returning to the theoretical debates in the outset of the chapter, cultural constructionists affirmed the positive roles of "invented traditions," even though they are not "authentic," in shaping the identities of local communities. The empirical analysis in this chapter helps us to re-think and question the readily affirmation, not because they are unauthentic, but to put it simply, because the "invented tradition" catering to tourists has not been enjoyable for the local residents. Accordingly, what the locals tried to take back was not objectively authentic traditional culture, but "living traditional culture" which is based on co-subjective and emotional "aesthetic reality."

Notes

1 The content of the chapter is based on Adachi (2010, 114–161) with major revisions.
2 This fieldwork was mainly carried out between 1997 and 2005.

References

Adachi, Shigekazu. 2010. *Gujo Hachiman Dento wo Ikiru: Chiiki Shakai no Katari to Riariti* [Gujo Hachiman Living Traditions: Narratives and Realities of the Local Society]. Tokyo: Shinyō Sha.
Bruner, Edward M. 2005. *Culture on Tour: Ethnographies of Travel*. Chicago: University of Chicago Press.
Gujo Odori Hozon Kai. 1922–1946. *Taishō 11-nen Ikō Kaigi Roku* (Meeting Records Since 1922). Gujo Odori Hozon Kai. In Japanese.
Hobsbawm, Eric, and Terence Ranger. eds., 1983. *The Invention of Tradition*. Cambridge, UK: Cambridge University Press.
Inokuchi, Shōji. 1977. *Denshō to Sōzō: Minzoku Gaku no Me* [Succession and Invention: Eyes of Folk Culture Studies]. Tokyo: Kōbundō.
Torigoe, Hiroyuki. 1997. *Kankyō Shakaigaku No Riron to Jissen: Seikatsu Kankyō Shugi No Tachiba Kara* [Theory and Practice of Environmental Sociology: Perspectives of Life Environmentalism]. Tokyo: Yūhikaku.
Yanagita, Kunio. [1941] 1970. "Tanoshī Seikatsu" [Enjoyable Life]. In *Teihon Yanagita Kunio Shū: Dai 30 Kan* [Yanagita Kunio Standard Collection: Vol. 30], 187–202. Tokyo: Chikuma Shobō.

13 Embracing the Enemy's Legacy
Historical Environmental Preservation in Daegu, South Korea

Rie Matsui

From the end of the 19th century until Japan's defeat in World War II, Korea was under Japan's colonial rule. Even after liberation, South Korea and Japan have been unable to resolve issues originating from the colonial era. This inability is because the relationship between Japan and South Korea, which had been that of the colonizer and the colonized, was forced under the Cold War system to transition to a relationship between "friendly nations" that belonged to the US-led capitalist camp (Kim 2014). As a result of this shift, the history of colonial rule was slowly but steadily forgotten in the consciousness of the Japanese. By contrast, in South Korea, colonialism is far from a thing of the past, and in turn most vestiges of colonial rule have been actively eliminated.

As definite and tangible vestiges of colonial rule, colonial-era structures remaining in South Korea were targets for historical reckoning after the liberation of Korea. Though many such colonial-era structures (e.g., Shinto shrines and colonial government buildings) were demolished, some continued to be used for various reasons. This chapter analyzes how one community in particular navigated this process of historical reckoning, continued practical use, and local meaning associated with colonial-era buildings.

In this chapter I focus on a local community in Deagu where the community did not destroy distinctive, Japanese-style houses called *Jeoksan-Kaok* ("enemy's houses"); rather, it is actively preserving them. What led the community to preserve these houses? In order to account for the logic of preservation, I draw on the perspective of life-environmentalism, focusing on the residents' historical everyday life experiences.

Historical Environment Preservation in Postcolonial Cities

In *Of Planting and Planning: The Making of British Colonial Cities* (1997), Robert Home examines how colonialism shaped the development of colonized cities through three contemporaneous ideologies: state control, capitalist, and utopian ideologies. Home's work, along with some other postcolonial research on the history of urban planning, focuses on the ways in which these cities were formed at the time of colonization. Accordingly, current residents in formally colonized cities are typically portrayed as passive actors whose lives continue to be shaped by the legacy of colonial urban planning. Such a perspective, however, is not helpful in answering why a local community would actively preserve the *Jeoksan-Kaok*, tangible vestiges of colonial rule that were targets for systematic elimination after the liberation of Korea. We need instead to start with a different view, one that focuses on the active agency of the residents of postcolonial cities in preserving the *Jeoksan-Kaok*.

To that end, life-environmentalist work on "historic environment" offers a useful point of reference. For example, Torigoe (2018) provides an account of Taketomi Island, a small island in Okinawa Prefecture of Japan, which is well known as a tourist destination for its streetscapes of red-colored tiled roof buildings and sandy streets. However, those red-roofed streetscapes did not historically exist throughout the Island as they do today. Red-titled roofs first appeared only in the late Meiji period, and only wealthy households could afford those roofs at that time. A widespread use of red-titled roofs began only in the 1970s. It is important to note, however, that these roofs were not installed simply to attract tourists. Rather, as Torigoe posits, the preservation and further installation of red-tiled roofs must be understood as the result of local activism to overcome their economic hardship, without excessively relying on external capital, as well as to push back the discrimination and prejudice with which they are viewed by outsides (e.g., those from the main Okinawa Island, and of mainland Japan). In this way, Torigoe interprets the historic environmental preservation as the Taketomi island community's effort to live well, not just materially, but also with the sense of pride and dignity.

This chapter applies the life-environmentalist perspective to understand the *Jeoksan-Kaok* not merely as vestiges of colonial rules, but as a built environment that has long supported people's livelihoods after liberation. In doing so, I articulate the historically formed local logic behind the preservation of the *Jeoksan-Kaok*.

Overview of the Case Study Area

The community which is the focus of this chapter is found in Daegu Metropolitan City (Figure 13.1). Located inland in southeastern Korea, Daegu has a population of approximately 2.4 million people. As the third largest city in South Korea after Seoul and Busan, Daegu has been a key city in the southeastern Korean Peninsula since the 14th century. The community in question is situated in the Bukseongro neighborhood, located in the Jung ("central") District of Daegu. The name Bukseongro translates to "the north castle wall road."

The history of Bukseongro is intimately connected with Japanese colonial expansion, which began in the late 19th century. The first Japanese colonists arrived in Daegu in 1893, 17 years before the Japanese annexation of Korea. At the time, Daegu was a walled city surrounded by the Daegu Fortress. However, the fortress was demolished in 1907 by the demand of the Japanese Settlement Corporation in Daegu. Part of the former fortress site became a street; a busy, bustling thoroughfare connecting Daegu Station and Daegu Shrine, the

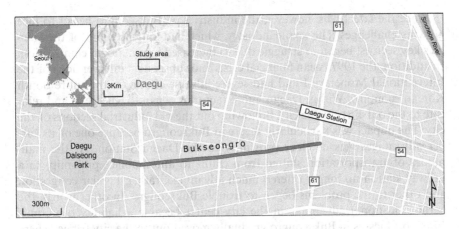

Figure 13.1 Bukseongro neighborhood in Daegu. Map by the author.

two most important sites for the Japanese colonists. This newly built street became Bukseongro. Many colonists moved in and built houses lining both sides of this street. As a result, the fortress symbolizing the precolonial spatial order in Korea was destroyed, and was replaced by a Japanese-style streetscape reflecting the new colonial order.

After Japan was defeated in World War II in 1945, the colonists were repatriated from the Korean Peninsula. The structures which had been built and owned by the Japanese colonists were passed into the hands of the Korean people. It is important to note that, at the time, people had no option but to use the buildings which were readily available to them. The structures built by the Japanese colonists—in particular the *Jeoksan-Kaok*—were suited to the humid Japanese climate, and therefore not particularly suitable for the drier, and colder winter climate in the Korean Peninsula. However, with little choice of housing available, the new inhabitants made do, first removing clearly Japanese elements such as signs written in Japanese, concealing the exterior with fencing, and adding new large signs. Subsequently, some of the colonial-era structures were repeatedly repaired and renovated for more comfort and ease of use, and have survived to this day.

Today Bukseongro is known for a large number of tools and machinery businesses and workshops, clustered along the street. The origin of the industrial town can be traced to the Korean War, which broke out in 1950, soon after liberation from the Japanese colonial rule. Many parts of the Korean Peninsula were devastated during the war, but Daegu was largely spared from damage. A black market emerged for military supplies such as tools because of its proximity to the American army units stationed near Bukseongro. The street stalls producing and selling those tools and machineries eventually laid the foundation of the long-standing industrial cluster in Bukseongro. As one of the industrial centers of Daegu, Bukseongro's local economy reached its peak during the 1980s and 1990s.

Since the late 1990s, however, the industries of Bukseongro began to face growing challenges. The textile industry, which had long been one of the Daegu's key sectors, began to struggle due to increased international competition. In addition, the 1997 Asian financial crisis and subsequent interventions of the International Monetary Fund had severe adverse effects on the entire South Korean economy. Within Daegu, more factories began to emerge in and relocate to the outskirts of the city, hollowing out the old industrial sections of the city. The impact was particularly severe in Bukseongro because one of its earlier advantages was its close proximity to the Daegu Station. As the main modes of transportation shifted from trains to automobiles, proximity to a railway station no longer offered much advantage; rather, regular congestion and other traffic problems began to trouble Bukseongro. In other words, the earlier locational advantage in the city center became a liability over time. Many businesses in Bukseongro gradually moved out to the city fringe, where a new business park was built.

In Bukseongro, the remaining colonial-era buildings continued to be lived in and used by the people who stayed. However, the economic downturns and the relocation of the businesses to the suburbs led to an increasing number of empty stores in Bukseongro, and the colonial-era buildings began to deteriorate rapidly. In this way, the streetscape of Bukseongro changed, reflecting the eroding vitality of the community that once flourished as the region's industrial hub.

"Rediscover Daegu" Citizen's Movement

As the Bukseongro area began to lose its earlier vibrancy, a citizen's movement was founded in Daegu in the early 2000s. The movement, called "Rediscovering Daegu," involved citizens studying the city's history, focusing on the remaining buildings and sharing the findings with fellow residents (Matsui 2008). This movement originated as the activities of the Daegu YMCA in 2001. It involved college student volunteers who sought out and made an inventory of cultural resources of the city, mapped the resources, and offered local tours for visitors based on the outcomes. These activities led to the creation of a nongovernmental organization (NGO), Citizen's Solidarity for Ecological Street (CSES), in 2002.

Bukseongro became one of the focal points of CSES because the area embodies the visual legacy of Daegu's modernization process. CSES itself went on hiatus around 2008, but other NGOs succeeded its activities. Those NGOs include the Time and Space Research Institute (established in 2012) and HOOLA (established in 2016), both of which focus their activities in Bukseongro. These NGOs expanded their activities through their partnerships with governments and research institutes.

The critical point here is that the citizen's movement had to *rediscover* Daegu. As stated above, the vestiges of Japanese colonial rule in South Korea were actively eliminated after its liberation. This meant that the narratives of modern history of Korea from the late 19th to the 20th centuries came to center around "the history of resistance to the Japanese colonial rule." As a result, narratives of modern history which did not fit into this framework tended to be neglected and forgotten. In addition, modern historical narratives often focus on national, macro-level events such as the Korean War, rapid economic growth, and the prodemocracy movement. Against this background, micro-level, everyday experiences of people who lived under Japanese colonial rule were forgotten in South Korean society. This explains why the local history of Daegu, other than that of resistance to the Japanese colonial rule, needed to be rediscovered through the citizen's movements.

The Rediscover Daegu movement, based on extensive interviews by citizen researchers, shed light on the history of the city under colonial rule, with particular focus on the daily lives of common people. Their grassroots fieldwork culminated into a revised modern local history of Daegu that is distinct from the history of the nation state. Another significant feature of the movement was that it not only focused on the colonial period, but also paid great attention to how people lived after liberation. Those NGOs, which succeeded CSES and worked closely with the Bukseongro community, conducted comprehensive field surveys in the Jung District, which includes Bukseongro. They studied the current conditions of traditional Korean houses, called *hanok*, as well as those of other historical buildings, and interviewed business owners and inhabitants in the area. Those interviews covered such topics as life histories of the inhabitants in Bukseongro, and various specialized knowledge and skills of workers in the neighborhood.

One of the important findings of the study was the distinct character of the Bukseongro community, firmly grounded in the shared life experiences of the residents and workers. For example, memories of living under colonial rule were widely shared among the long-standing residents and workers of the Bukseongro community. Those stories, which are not solely about the resistance to the Japanese colonial oppression, were then passed on to those who were born in the postliberation period.[1] Such traces of the historical continuity can also be seen in the fact that some technical terms in local workshops are either in or derived from the Japanese language.[2] Thus, the history of Bukseongro under colonial rule is remembered and passed on not merely by individuals, but by the community.

The character of the Bukseongro community is also evident in stories from the postliberation era. There is a saying, "all problems can be solved in Bukseongro." This is because of a large array of materials and tools that are available for customers to choose from in Bukseongro, and because of the presence of craftsmen with a wide range of skills and techniques who can make whatever products customers need. Not only that, knowledge about where different materials, tools, craftsmen, and skills/services can be found was also shared throughout the Bukseongro community. Thus, for example, if one wishes to remodel their bathroom and goes to a store that sells bathroom sinks, and other interior equipment, not only they help design the interior and identify necessary materials, but they can also refer the customer to a metal processing factory on the street that can make customized interior and exterior parts. The above phrase therefore encapsulates the nature of the Bukseongro community, not only where a wide range of resources are available, but also where a tight-knit community shares the resources and information.

Needless to say, requirements for manufacturing industries have been constantly changing as industrialization progresses over time. This means that workers in Bukseongro also needed to adapt to the changing time, and they have done so by sharing the latest information and helping each other with skill development. Many businesses on Bukseongro are one-person companies. Presidents (who are also craftsmen/engineers) of these companies often get together over lunch or drinks, and talk about latest technologies and business information (Matsui 2022). In short, Bukseongro has been able to thrive as an important industrial hub by building a community, which is characterized by strong bonds among the residents and workers.

The Rediscover Daegu movement was an important achievement in two ways. One was that it shed light on the local history of the city under colonial rule, which is more complex than the conventional national-scale narratives. Another is that the movement elucidated the daily lives of the people who lived through the colonial and postliberation periods on a continuous spectrum. By doing so, it also articulated the unique character of the Bukseongro community that was formed through the life experiences of the people.

To be sure, the movement, which was led primarily by "outsiders," did not have unanimous support from the Bukseongro community in the beginning. Some residents and workers reacted supportively, if not enthusiastically, narrating their personal experiences and sharing historical documents that they had. Others showed little interest in the movement. In addition, those community members who had experienced economic prosperity during the 1980s and 1990s lament the area's comparative decline in recent years. Of course there are still new businesses and workers that come to Bukseongro, and the area has not completely lost its role as an industrial hub. Yet, without question, there is definitely a gap in perception toward current Bukseongro between long-term members of the community, seeing "struggling Bukseongro," and members of the citizen movement, who see "Bukseongro full of possibilities" and are aspired to work with the members of the local community.

After all, the citizen movement was not meant to bring back the past economic glory. Rather, it was to accept and affirm the history of the community as it is. This basic stance resonated with the consciousness of a large number of Bukseongro community members. It is in this context in which the significance of restoring the *Jeoksan-Kaok*—enemy's houses—was articulated.

Life Stories Visualized: The Historical Architecture Restoration Project

In Bukseongro many buildings from the colonial period have been used as tool and machinery stores, and as workshops and small factories. These buildings were often shrouded in a variety of wall materials and with large signs, so that one could easily discern the original forms of the buildings from their façades. As these buildings became vacant and their conditions began to deteriorate, a program was launched to rehabilitate the empty old structures: the Bukseongro Historical Architecture Restoration Project (Matsui 2017). One of the key features of this project was its emphasis less on the historical architectural values of these buildings, and more on the local historical experiences encapsulated in these buildings based on the results of interviews and field surveys. Accordingly, the project prioritized the restoration of buildings that were considered to embody and represented the historical experience of the Bukseongro community.

The restoration project began with a private sector–driven pilot project, targeting four buildings between 2011 and 2013. Restoration typically involved removing the elements added to a building which masked its original form (such as Korean signs, modern cladding, and metal roller doors), revealing its original features (such as wooden window framing, wooden cladding, and tiled roofs), replacing missing elements in the original style (such as stucco wall finishings), and preserving the original appearance by applying modern exterior coatings which showcase its beauty. As a result of the pilot project, deteriorated vacant buildings were reimagined into a cafe, art gallery, architectural studio, and guest house. Importantly, the four restored buildings were all *Jeoksan-Kaok* built during the colonial era. The overall restoration project was overseen by architects, but local tool and machinery store staff as well as craftsmen of small factories played essential roles in specific aspects of the restoration process, such as procuring and custom-building interior and exterior elements. The pilot project was subsequently developed into a public-private partnership project for urban landscape design, which involved the restoration of historical buildings built before the 1960s. Between 2014 and 2016, approximately 30 buildings were restored and put into new and innovative uses under the program.

Through these restoration projects, historical buildings, which, despite always being there, had been rendered invisible by later additions and coverings, began to reappear one by one in Bukseongro. In particular, the *Jeoksan-Kaok* buildings, among other restored buildings, have a conspicuous presence in the eyes of visitors to present-day Bukseongro. However, it is critical to understand that the *Jeoksan-Kaok* had always been a familiar presence for the people of Bukseongro. That is to say, the *Jeoksan-Kaok* of Bukseongro were not preserved for their value as vestiges of the colonial era based on the state history; rather, they were preserved as the built environment which had supported the everyday lives of the local people. One of the symbolic illustrations is the Bukseongro tool museum (Figure 13.2), which preserves the colonial Japanese structure outside, and exhibits a number of tools produced during the early postliberation period inside (Matsui 2017). In other words, the restoration of the *Jeoksan-Kaok* was accepted in Bukseongro precisely because the goals of the project were articulated through the locally grounded, concrete life experiences of the community members.

Figure 13.2 Bukseongro tool museum. Photograph by the author.

These restoration projects can be considered an endeavor to "visualize life histories" (Matsui 2018), in which the streetscape with restored buildings visually encapsulates the Bukseongro community's life experiences to date. Thus, the *Jeoksan-Kaok* are an integral part of the history of Bukseongro and bring the community's past into the forefront of the residents' consciousness. In turn, they provide opportunities to reflect on the present and to envision the future of the community as an extension of its past.

Conclusions

One of the goals of the postcolonial studies is to help create a genuinely free society by exposing and critically scrutinizing the legacy of colonialism. Studies on colonial cities referred to in this chapter, such as Home (1997), follow this intellectual tradition. While offering important insights, some of these existing studies tend to assume the presence of colonial-era buildings as the evidence of persisting effects of colonialism on current residents of the city. This chapter instead calls to focus more explicitly on active agency of local actors in postcolonial cities in shaping their living environment, and to learn from their concrete, everyday activities and experiences that continue from the past to the present.

This chapter has examined the preservation and restoration of the *Jeoksan-Kaok* in Bukseongro, which, from the perspective of the national history, would be a target for thorough elimination because they are tangible vestiges of colonial rule. Rather than casting this as the evidence of persisting colonial effects, however, I interpret the preservation of the *Jeoksan-Kaok* as a manifestation of the active agency of the residents of the city. The approach adopted in this chapter is informed by life-environmentalism, which emphasizes the roles of historically accumulated life experiences in understanding the nature of the present-day urban landscape.

Let me return to the original question of the chapter: Why were the *Jeoksan-Kaok* preserved in Bukseongro? That is, why did the *Jeoksan-Kaok* change from the subject of elimination to that of preservation? In postliberation South Korea, creating a new nation state was certainly an urgent task, and erasing the traces of Japanese colonial legacy was critically needed. At the same time, as this chapter has revealed, it was not practical or feasible to completely eliminate the all vestiges of colonialism for people to carry out their everyday lives. Thus, by living and working in remnants of the colonial era, new communities emerged in the country. Bukseongro, a busy commercial street during the colonial era, prospered as an industrial center where "all problems can be solved" in the postliberation period. And that prosperity was supported by the dense network of people, many of who lived and worked in the *Jeoksan-Kaok*, forming the postliberation Bukseongro community. The accumulation of lived experiences, not some ideological forces or objective architectural values, are one of the central reasons why the preservation of the *Jeoksan-Kaok* was accepted as legitimate in the eyes of the local people of the Bukseongro community.

Another reason is the presence of the citizen's movement, which rearticulated the local history of Daegu by focusing on the daily lives of the people who lived through the colonial and postliberation periods on a continuous time horizon. By not segmenting historical time periods (e.g., pre-, during, and postcolonial periods), the citizen's movement was able to identify the historically formed, unique character of the Bukseongro community that is rooted in the lived experiences of the residents and workers.

For outsiders, the preservation of the *Jeoksan-Kaok* may be interpreted as a strategy to revive the economy of Bukseongro, perhaps by attracting tourists, which has lost its vibrancy as an industrial hub. The preserved and restored historical buildings may indeed bring in tourism or other forms of economic benefits. Nevertheless, as this chapter has shown, the most important underlying rationale behind the preservation project, facilitated by the citizen's movement and embraced by the Bukseongro community, was the positive affirmation of the community, including its turbulent history and the shared life experiences of its residents and workers.

Acknowledgment

This article was translated into English by Glen McCabe. The author would like to take this opportunity to thank him for his assistance.

Notes

1 For example, when I visited Bukseongro with my students in 2019, a person in his 60s approached us, and began talking to us what was at the site of the current parking lot. According to him, who lived by the site until he was 15 or 16 years old, there was a Japanese department store during the colonial period, and there was an elevator in the building, which was very uncommon at that time. After the Japanese pulled out, the U.S. military came and used the building as a tax office, after which it was torn down.
2 Examples include *kireppashi* (a piece of wood or metal left over after processing) and *nakama* (a broker who buy tools in other places and sell them to retailers on Bukseongro).

References

Home, Robert K. 1997. *Of Planting and Planning: The Making of British Colonial Cities*. London: Spon.

Kim, Sungmin. 2014. *Sengo Kankoku to Nihon Bunka - "Washoku" Kinshi kara "Kanryu" made* [Post-war South Korean and Japanese Popular Culture: From the Prohibition of the "Darned-Japanesque" to the "Korean Wave"]. Tokyo: Iwanami Shoten.

Matsui, Rie. 2008. Kankoku ni okeru Nihon-shiki Kaoku Hozen no Ronri - Rekishiteki Kankyō no Sōshutsu to Chiiki Keisei [The Logic of the Historic Preservation Movement in the Context of Community Formation: A Case Study of CSES in Daegu, Korea]. *The Annual Review of Sociology*, 21: 119–30.

———. 2017. Keikan Hozen o Tsūjita Toshi no Keishō - Kankoku/Daegu no Kindai Kenchikubutsu Rinobēshon o Jirei Toshite [Urban regeneration Through Landscape Preservation Movements: From a Case Study of Renovation Planning of Historical Architecture in Daegu, South Korea]. *Contemporary Sociological Studies*, 30: 27–43.

———. 2018. Komyuniti ni Jeoksan-Kaok o Toriireru - Kankoku/Daegu no Bukseongro ni okeru Seikatsu Jissen no Rireki no Kashika [Incorporating the Jeoksan-Kaok into a Community: Visualizing the History of Life Practices in Bukseongro, Daegu, South Korea]. In *Seikatsu Kankyō Shugi no Komyuniti Bunseki [Community Analysis of Life-Environmentalism]*, edited by Hiroyuki Torigoe, Shigekazu Adachi, and Kiyoshi Kanebishi. 171–90. Kyoto: Minerva Publishing.

———. 2022. Machikōba to Komyuniti - Kankoku/Daegu no Bukseongro de Hataraku Gijutsusyatachi no Katari kara [The Links between Small Factories and a Local Community: A Case Study of the Narratives of Local Workers in the Bukseongro Neighborhood of Daegu City, Korea]. *Tourism and Community Studies*, 1: 85–97.

Torigoe, Hiroyuki. 2018. The Historic Environment in Opposition to Social Inequalities. In *Facing an Unequal World*, edited by Raquel Sosa Elizaga, 146–51. London: Sage.

14 Boxing Camp as a Community School

Local Boxers in Metro Manila, Philippines*

Tomonori Ishioka

Boxing Camp as a Total Institution?

Boxing gyms are found all over the world, but the vast majority of boxing *camps* are found in Southeast Asia, such as the Philippines and Thailand. Boxing camp is a space where boxers not only train but also live together. This is different from boxing gyms where boxers (including many professional ones) who have other jobs during the day come for training in the evening. After training, they return to their homes and apartments at night. Boxing gyms in many industrialized countries are a training space, not a living space. At boxing camps, on the other hand, boxers spend 24 hours a day with other fellow boxers. They all get up early in the morning, jog, have breakfast, wash and clean, train, hang out in the neighborhood, have dinner, watch TV, and go to bed at night.

In *Asylums* (1961), Goffman argues that the basic arrangements of modern society involve the separation of work, play, and sleep. This separation of activities is the reason that we commute to work and school from home, and is the result of rationalization of social life, as argued by Max Weber. Moreover, Goffman suggests that modern society also has spaces where the barriers separating different spheres of life are broken down. He calls such spaces a "total institution" whose examples include psychiatric hospitals, prisons, the military, and lodge camps. The word "total" suggests that "all aspects of life are conducted in the same place and under the same authority" (Goffman 1961, 17), which controls and disciplines the members. Inevitably, many of these total institutions are established as penal institutions. In this sense, boxing camps may be seen as a kind of total institution, in which members live, train, and are disciplined collectively, in order to achieve the goal of becoming successful boxers.

In the Global South, however, total institutions do not always simply provide collective discipline to their members; rather, they also often offer collective care

---

* The empirical sections of this chapter draw on Ishioka, Tomonori. 2012. "Boxing, Poverty, Foreseeability – an Ethnographic Account of Local Boxers in Metro Manila, Philippines." *Asia Pacific Journal of Sport and Social Science* 1 (2–3): 143–55, and are used with permission.

DOI: 10.4324/9781003185031-18

to marginalized populations. In other words, total institutions embody not only the logic of control and discipline, but also the logic of safeguarding livelihoods. For example, Kitiarsa (2013), based on his study in the northeastern province of Thailand, shows that Buddhist temples provide poor young men from rural areas with the basis to live independently. In a similar manner boxing camps in Manila can also be said to embody such a function of offering care in addition to that of discipline and training.

To articulate this point further, Hiroyuki Torigoe's classification of two types of education is useful. Following the works of Kunio Yanagita, Torigoe (2008) suggests that "extraordinary education" (*hibon kyōiku*) is the kind of education that helps one to "become different from others." Examples include teaching the youth how to speak English fluently, or to become top athletes. On the other hand, "ordinary education" (*heibon kyōiku*) teaches members to become, or to live, like other community members. Examples include learning to swim in coastal fishing communities, and learning basic knowledge about planting and harvesting in farming villages. Moreover, "ordinary education" extends to the acquisition of everyday skills to live well socially, including learning how to greet and interact with people, how to follow regular routines of everyday life, and how to mutually aid each other (Kropotkin 1902).

I argue that boxing camps in an urban neighborhood of Manila indeed embody both of these functions of total institution: training to become champions and earn a fame (extraordinary education) and learning to secure "normal everyday life" in an urban neighborhood (ordinary education). What this means is that total institutions in the Global South, especially in Southeast Asia, are not only institutions for the state to manage its marginalized populations (Foucault 2007), but are also a local institution, or what I call "community schools," to help members cope with everyday problems arising from poverty and marginalization.

The study is based on an ongoing research from 2002 to present in an urban neighborhood of Manila, including live-in ethnographic research at Boxing Gym E in Paranaque City between April 2005 and March 2006. I accompanied the gym members on their early-morning road-training, and trained with them during their afternoon gym routines. The live-in ethnography has been followed by a series of shorter-term follow-up studies spanning between two weeks and a month at the gym on an annual basis until now.

Where Do Boxers Come From?

Gym E is located alongside Sucat Road, which crosses Paranaque City. Across from the gym is Metro Manila's famous cemetery park, Manila Memorial, where the boxers carry out their daily morning routines. The gym is situated in the midst of a busy market; one can hear the sound of jeepney engines and radios blaring from neighboring houses before 6 am. Despite the chaos, however, the boxers diligently perform their daily morning routines.

The daily lives of the boxers in Gym E are deeply connected with the world of poverty. The Fourth Estate area has over 6,000 squatter households. One in every three to five people living in Metro Manila is a squatter resident. However, this does not mean that all squatters live in poverty. There are also squatter households that live on remittances from a household member who has pushed to work in countries such as Saudi Arabia (employment in factories subcontracting within the oil industry being typical for Filipino men) or Singapore (employment as domestic staff being typical for Filipina women). Such households enjoy a higher standard of living than other squatter residents. On the other hand, amid the context of the economic globalization since early 2000s, there is an increasing number of squatters living in previously unseen levels of poverty. In this way, the squatters of Metro Manila are facing class polarization.

While most active boxers live in the gym (locally expressed as "staying in"), those with families or retired boxers live elsewhere ("staying out"), often in the squatter area (Figure 14.1). Due to lack of proper water and sewage facilities, the residents have to purchase water from the neighboring areas and carry it back to their houses, discharging wastewater into open ground. Moreover, it is common for the area to become inundated during heavy rains. The only consolation is the inexpensive rent of 1,000–2,000 pesos per month.

As part of the squatter area culture, the boxers buy the ingredients for their morning and evening meals at the Fourth Estate market and prepare and eat these meals together. The boxers can afford to buy fruit juice, sweets, and prepaid cards for their mobile phones, despite their small allowances, because these goods are relatively inexpensive. In some cases, they buy items on credit

Figure 14.1 Squatter area near the boxing gym. Photograph by the author.

and pay it off with the "fight money" earned from a boxing match. Boxing is a highly regarded profession in the squatter area and has served to strengthen ties between the boxers and the residents. In fact, even those involved in gang activities and drug peddling in the neighborhood show respect to the boxers. It seems that the gang members particularly respect the boxers' lifestyles and their single-minded commitment to winning in the ring without any regard for social status or material well-being. In addition, many squatter residents call out to the boxers to engage them in conversation about their next match or their match on TV the previous day. This suggests strong cultural ties between the squatter residents and the boxers.

Unemployment and poverty are the biggest social problems in Fourth Estate. In Tagalog, the unemployed are known as "istambay," which literally means "those on standby." However, in recent times, the word has come to refer specifically to the unemployed who are the product of patterns of poverty in the Philippines. A resident of Fourth Estate, unemployed, says, "We capable of working, but there aren't any jobs in the Philippines. There aren't any jobs for us poor people." The word "istambay" can also be used to describe a situation in which even those who are desperate to work are unable to find jobs.

As is the case for other members of boxing gyms in the Philippines, most of Gym E's boxers live on a "stay in" basis. All boxing members of Gym E are professional and make a living solely through boxing. As mentioned above, boxers who "stay out" are mostly those who have a family, and even these boxers make a living through his income, or their managers provide for their daily necessities. Many boxers in heavily industrialized countries tend to maintain part-time jobs as well, whereas in the Philippines, boxers are true to the profession of boxing. Thus, boxing camps/gyms in the Philippines have to take care of all aspects of boxers' daily lives and, as a result, in a sense have control over their boxers' existence.

We can broadly divide the boxers at Gym E into those who left their hometowns to pursue the sport in Metro Manila, and those who lost their jobs in Metro Manila and joined the gym due to their unemployed status. To take the example of the 47 boxers who were members of the gym between November 2002 and September 2008, 27 left their hometowns to become boxers, 14 were already in Metro Manila and searching for a job when they turned to boxing, and the remaining 6 were boxers who had transferred from a local gym. The professions of those who came to Metro Manila in search of work prior to boxing varied. They included butchers, fishermen, construction workers, bartenders, waste collectors, street-side hawkers, and Jeepney drivers. Some even confessed to having robbed nightclub patrons (in many cases, foreign nationals) by threatening them with a knife, because they lived on the street and needed money to survive. There were also some boxers who stated that they became boxers simply because they were constantly starving.

In terms of their origins, most of the boxers come from the southern areas of the central region of the Philippines, particularly eastern Visaya on

Leyte Island. One possible reason for this is that the manager of Gym E is friends with boxing instructors on Leyte Island. Other birthplaces include Bicol, central Visaya and Soccsksargen, which are generally more economically impoverished than the capital region. With regard to the educational backgrounds of the boxers, of the 47 boxers at Gym E, 31 were secondary-school dropouts, nine had completed their secondary education and seven had completed only their primary education. None of the boxers had gone to university, junior college, or vocational school. Even though cultural factors (e.g., public respect for boxers) play an important role, these socioeconomic conditions are undeniably critical factors in drawing young men to boxing.

As for the boxers' ages and the number of years they had spent at Gym E, the youngest boxer was 15 and the oldest was 34. Most of the boxers were between the ages of 17 and 19, and had begun their boxing careers at Gym E. The boxers train and build their experience while competing in matches. However, most of them give up boxing within two to four years. During the course of my surveys, 26 of the 47 boxers retired from boxing within two years. This tendency for boxers to quit after a relatively short time is due to the physical demands of the sport. There were also those who were enthusiastic, yet did not have the right qualities to be a boxer. The gym members also included a 34-year-old boxer who was a former Philippines national champion; he was said to have all the qualities required for the sport, which is what had allowed him to continue boxing at his age. However, of the 47 boxers, only two were over the age of 25. That is, over 95% of boxers are 25 or younger. The majority of boxers at Gym E enter the world of professional boxing in their late teens and retire by the time they reach 25, with many quitting their boxing careers even sooner.

Over the course of their careers, boxers progress through the rankings of Four Rounder, Six Rounder, Eight Rounder, and finally Ten Rounder (some Ten Rounders are selected to participate in title fights). However, among the 47 boxers at Gym E, only 10 advanced to the ranking of Ten Rounder, indicating that 80% of those who take up boxing retire without ever having received the rank of Ten Rounder, let alone becoming champions.

If we consider boxing camps as the total institution solely on the basis of "extraordinary education," then Gym E fails many of its members. It has not significantly advanced social mobility of its members, which is often portrayed as the key nexus of poverty and sports (Carrington 1986; Spaaij 2009; Singer et al. 2011). In fact, the fact that over 80% of boxers retire before reaching the rank of a Ten Rounder may indicate that the boxers themselves do not see the sport solely as a way of gaining social mobility. This is the reason that understanding boxing camps also as the institution of "ordinary education" to live their everyday life well, and to be like other members of the urban community may be a viable interpretation. To that end, let us look more closely at the everyday life of the boxers at Gym E.

Entering into the Everyday Life of Local Boxers

Try-Out

To become a boxer at Gym E, candidates must go through a form of sparring called a try-out. Those who wish to participate in a try-out go to the gym at 1 pm on Mondays and Fridays. No reservation is required, and anyone is allowed to take part as long as they are at the gym at the appointed time. As a prerequisite, they need to inform the gym staff of their intention to undertake a try-out. Those who perform well during the try-out are admitted into Gym E as boxers and are allowed to reside in the gym. The number of candidates varies each day: although on some days there are no try-out candidates, on other occasions there are five or more.

The try-out candidates' sparring partners are either professional boxers who are soon to make their debut or Four Rounders who have recently made their debut. There is a major gulf between the abilities of these professional boxers who have been training for close to a year and the try-out candidates who have had very little experience of boxing. Thus, in most cases, the three-round try-out ends in the first or second round by a knockout. During try-outs, the trainers judge the potential of the candidates and decide whether they can make the grade. At the same time, they ascertain whether the candidate has the requisite "strength of heart" to become a boxer.

I have attended try-outs on several occasions, but have yet to see a member become a boxer at Gym E by clearing try-outs in a single attempt. While candidates are immediately told by the trainer whether they have passed the try-outs or not, in normal circumstances the boxers who have passed the first try-out must undertake a second one. In the majority of the cases, those who pass the second try-out are allowed to join Gym E. However, in some instances, they may be told to undergo an additional third try-out as a final check.

Daily Routine

Those joining Gym E lead communal lives centered on the gym. Boxers have a strict timetable. A typical day at the gym consists of morning routines from 6:00 am, breakfast, training again from 1:00 pm, and then dinner, which is generally eaten with their manager.

The boxers wake up every morning at 5:30. Once they are up, and have brushed their teeth, washed and changed, they go for a 3-kilometer run at 6:00 am, and then practice shadow-boxing. Because the timing of early-morning gym routines differs for each boxer, those who finish first can return to their rooms to take a shower. Those who do not take a shower are labeled as *mabaho* ("stinky") or *walang ligo* ("no shower") by other boxers and their manager.

At around 8:00 am, after all the boxers have taken a shower, they proceed as a group to the local market to buy ingredients for the day's meals. The reason they go as a group is because this allows each boxer to buy his own water and bread for snacking. The gym chooses those responsible for preparing meals, with the cooks at the time of the study being Alan, a Six Rounder, and Nick, a Four Rounder. These two members receive money from their manager to prepare breakfast, which consists of soup and white rice. The boxers eat breakfast at around 9:00 am.

Following breakfast, each boxer returns to his room. The period between then and 1:00 pm is allocated as free time. The boxers spend this time in a variety of ways, such as watching television, doing their laundry, or visiting friends in the neighboring squatter area. The boxers' training recommences at 1:00 pm. They arrive at the gym's training area and begin by wrapping their hands, followed by warm-up exercises in front of a mirror, after which they begin the various activities listed on their training schedules. The trainers come to the gym at around the same time, interact with the boxers while they prepare for training and then discuss their training schedules for the day. In the case of Gym E, Mondays, Wednesdays, and Fridays are largely geared toward sparring. However, boxers can focus more on their own training programs in consultation with their trainers on Tuesdays and Thursdays. Saturdays are set aside for weight training, which is why they do not practice techniques on that day.

Once training begins, boxers do three to four rounds of warm-up exercises and shadow-boxing. When they complete these activities, they engage in sandbag and mitt training, followed by punch ball, speedball, more shadow-boxing, and then skipping. Finally, they stretch and do more weight training, which marks the end of training for the day. Training finishes at around 2:30–3:00 pm. Thus, on average they spend less than two hours training per day. During that time, the trainers observe the boxers' every move and accordingly give them guidance. The relationship between the trainer and boxer is fixed. Among the three trainers at Gym E, Sonny trains five boxers a day, Jun trains four, and Vinvin trains three. Trainers generally feel possessive of the boxers who are under their exclusive care, referring to them as "my boxers." During training, the trainers go from one boxer to another, giving instructions, acting as a partner for a boxer's mitt training or supervising another boxer's bag training. Trainers do not live in the gym like the boxers, but typically live in the neighboring squatter area. Trainers have complete responsibility for all aspects of their boxers' lives; for example, any boxers found to have been out late on the streets are severely reprimanded by their trainers.

Once training has finished, the boxers again have free time until dinner. They congregate at around 6:30 pm in the dining room, and everyone eats at around 7:00 pm (Figure 14.2). At dinner, the manager is always present to share a meal with the boxers. This serves as an important time for the manager and boxers to interact with each other and talk about the day's events. Three long desks are put together, and the manager sits on what is called the "birthday seat." The boxers sit on both sides so that they surround the manager, and one of the boxers always says grace. Once all present have humbly made the sign of the cross, the meal is ready to commence. Dinner is not prepared by the boxers but consists of meals specially prepared for them at a restaurant next to

Figure 14.2 Dining room in the gym. Photograph by the manager of Gym E.

the gym which is run by the manager. A typical meal consists of grilled fish on white rice with soup. The manager is served a separate meat or fish dish made especially for him, and while the boxers all drink water, the manager has iced tea or soda. Once the meal is over, the boxers stay seated and chat for a while. I observed that they talked about a range of topics, such as the day's training, television programs, or their hometowns. The boxers do not leave the table until the manager says, "Okay, go back to your rooms;" having taken their dishes to the washing area, they then return to their rooms. After dinner, the boxers have free time until they turn their lights off and go to sleep, generally at around 10:00 pm.

Competing in Matches

Boxers follow practically the same routine from the time they join to their retirement. However, apart from this everyday cycle, boxers also have an additional cycle which encompasses the period from when a match is decided upon to when they appear in the ring.

Once a boxing match is set up, the boxers enter a period of intensive training. Unlike major matches such as a world title fight or even a Philippines title fight, local matches are usually set up between a week and one month prior to when the match is due to take place. Boxers who have been chosen to fight must lose sufficient weight to meet the weight limits for the match. Up until one week before the fight, the boxers lose weight gradually; during the last three days, they lose even more by going on an extremely stringent diet. These strict weight-loss regimes allow boxers not only to lose body fat but also to attain "social weight loss." Here, social weight loss refers to the act of boxers shedding all social desires that are not connected with boxing through suppressing their appetites and sexual desires and refraining from social activities outside the gym, such as visiting friends in the squatter area.

In fact, the act of shedding all social desires is also a standard that is upheld on a daily basis within the gym. However, prior to a match, this standard serves to constrain the boxers and is more strictly enforced. In other words, they engage in the act of shedding all social desires of their own "free will." Boxers are banned from having sexual intercourse and must refrain from masturbation. They eat meals that are largely free from carbohydrates, with the menu consisting largely of soup. They must go to bed at 9:00 pm, but get up and go jogging at 5:30 am wearing a sauna suit. I observed that as boxers get closer to the match, they are more focused and laugh less demonstratively and noticeably than on normal occasions. Even boxers who are generally witty and jovial isolate themselves in the week leading up to a match. They spend all their time in the training area or in their rooms, and do not visit the area surrounding the gym except during their routines. It is through this self-imposed control over all aspects of their social desires that boxers confront their matches.

Boxers at Gym E have an average of three matches a year, and these are possibly the only times at which they earn an income. Fight money is awarded for taking part in matches and is received regardless of whether the boxer wins or loses. The average fight money in the Philippines is 1,000 pesos (about US$20) per round. Thus, boxers receive 4,000 pesos from a match consisting of four rounds, or 10,000 from one consisting of ten rounds. Of the fight money, 45% goes to the manager and 10% to the boxer's trainer, which means that the boxer receives only 45%. Therefore, even if the boxer fights for 10 rounds, he receives only 4,500 pesos. However, if Ten Rounder boxers have the opportunity to compete in matches overseas, such as in Thailand or Japan, the market rate for fight money instantly increases. In the case of overseas matches, the average for one round is US$100; thus, a boxer lasting 10 rounds would receive US$1,000. Of this, the boxer would receive US$450, five times as much as he would earn in a domestic match in the Philippines. New boxers who have recently joined the gym look up to senior boxers who have experience of competing overseas, aspiring to the day when they too will have the opportunity to compete abroad.

However, it should be noted that even boxers competing at international level have an extremely low income. For example, Roselito, one of Gym E's top boxers, fought in three matches in 2005 (in Thailand, Japan, and the Philippines); however, his income for that year was only 39,200 pesos (about US$784).

This figure constitutes only 13% of the annual household income in Metro Manila calculated for that year, which was 300,304 pesos (about US$6,000). It is hard to get by on this amount even in the squatter area, which may explain why Roselito accumulated a large amount of debt in the form of store credit. Ultimately, although Roselito lived as a squatter in Fourth Estate, such things as eating at the gym enabled him to save money on food expenses, allowing him and his family to survive on the little they had.

Repetition, Long-Term Expectations, and a Collective Life

On average, it takes a new boxer five years from his debut to become a Ten Rounder. As mentioned above, only one in five boxers actually makes it to this level. Nevertheless, new boxers devote themselves to daily training with the dream of one day becoming Ten Rounders. By winning their debut four-round match and subsequent matches, they can gradually move toward achieving this goal and even competing abroad. In other words, entering the world of boxing gives young Filipinos specific expectations for their futures at least for the following five years, the time required for them to become a Ten Rounder. This applies to all gym members, which is why we are justified in describing this as a collective expectation. Becoming a boxer is equivalent to becoming a part of this collective expectation. Furthermore, it is important to note the acquisition of this collective expectation in parallel with the repetition which is a characteristic of boxing camps.

Since most boxers in the Philippines live on a "stay in" basis, they share all aspects of their lives with fellow boxers. "Repetition" is a crucial characteristic of shared everyday life, and has two cyclical rhythms. One is a repetition of time measured in units of single days and weeks. From waking until sleep, boxers turn physical training into an everyday affair through daily repetition of the same activities. In addition, the repetition of this routine from Monday to Saturday, and the designation of Sunday as a day off, serves to create a temporal order for each week. The other dimension is the repetition of time between matches, encompassing the period of time between a boxer entering the ring and his next match. Boxers are given a one-week holiday following their matches, after which they resume their daily training and strive to improve their skills until the next match is set up. Once their new match schedules have been determined, they once again begin the process of losing weight. Thus, until retirement, boxers constantly repeat this cycle of competing in matches, vacations, training, and weight loss.

Once these cyclical rhythms are established, boxers go on to work toward their long-term expectation of becoming a Ten Rounder. In addition to daily training, the new boxers plan their own specific futures by witnessing with their own eyes the process of adjustment that goes on between their senior boxers' matches. This should not be mistaken for the personal time of individual boxers; rather, it is collective time provided by the gym. In this sense, what we are presented with here is evidence of Durkheim's "collective life" (Durkheim 1965).

Becoming "Ordinary People" Through Foreseeability

To understand the sociological implications of boxers obtaining long-term expectations, we need to consider the poverty that defines the background of their daily lives. For young people from difficult socioeconomic backgrounds in the Philippines, taking up boxing certainly provides them with a possibility of attaining fame and fortune. Countries such as the Philippines have witnessed the rise of superstars such as Manny Pacquiao, and this has had a profound influence on the aspiration to become a boxer, earning millions of dollars a night, driving an expensive sports car, and living in an opulent house. However, this does not correctly describe the disposition of those impoverished young people who invest in boxing in places such as Gym E.

What becomes clear from everyday life at the gym is the acquisition of a regularized temporal frame, within which the boxers can live with the sense of "foreseeability," knowing what to do next, what happens next, and what can be expected of others (Ishioka 2012a, b). For the young men who came to the gym from impoverished rural areas, the sense of foreseeability was largely absent in their previous lives. The *istambay* who live in contemporary Manila's world of poverty, where unemployment and poverty are chronic, not only experience economic hardship but have also been robbed of foreseeability, which in turn has robbed them of their futures. Seeing this way we understand that "poverty" is not simply about the lack of jobs and income; rather, it also includes the deprivation of foreseeability. According to Gerald who retired from boxing:

> My life after retirement? I'm not busy at all—I am just kicking my heels. I have got plenty of time on my hands. However, it is meaningless time. I have got nothing to do; I am just idling my days away. When I was a boxer, I was busy and did not even have the time to hang out with my friends.... However, I was satisfied. It is good to have something to do. Look at the people living around here—those who do not have a daily routine are the most wretched of all.
>
> (Gerald, after his retirement)

Gerald's quote shows that an important role of the boxing camp is to its members a sense of foreseeability. Most boxers drop out of secondary education or less and come to boxing camps. They live a life of "nothing to do" and "not knowing what happens next" in the squatter area, but by starting boxing they have "things to do" and "know what to expect" in their everyday life.

Importantly, this is a collective space of learning; boxers spend their youth-to-adult years in boxing camps, and during this crucial time in their lives, they learn how to greet and talk with others, as well as the importance of mutual help by living together. Their daily routines and preparations for matches all teach them that they cannot live alone, which is also true in poor neighborhoods of Manila. In this sense, boxing camps are "community schools" teaching boxers how to become "ordinary adults" who know how to live well socially by supporting each other in everyday lives.

Conclusions

I began this chapter with the question of whether or not boxing camps in Global South can be understood as a total institution (Goffman 1961). My case study of Gym E in Manila shows that while it certainly embodies the logic of collective discipline, as typically seen in total institutions, its function goes beyond that, and extends to providing collective care. Here, "care" does not mean giving alms, but means to provide members with skills and knowledge to live a "normal" everyday life and the spirit of mutual help. In this sense, I argue that the notion of "community school" to cultivate "ordinary adults" (Torigoe 2008) is more appropriate in describing the nature of boxing camps in Manila. Importantly, members of Gym E do not only expect the gym to be a place to achieve their once-in-a-lifetime dream of becoming boxing champions, but they also seek, consciously or unconsciously, the sense of "foreseeability" where they can expect to regular, cyclical rhythms of everyday life. Indeed, conversely, we can discern that the sense of "poverty" does not only rise as the result of lack of income or material resources, but that it may also rise from the deprivation of foreseeability, not knowing what is coming up next or what they can expect from others.

In recent years, the view that squatter settlements, or "slums," in cities of Global South are mere "problems" to be swiftly eliminated have been increasingly under critical scrutiny (e.g., Matsuda 1998; Roy 2005). Even major international donor organizations like the World Bank now recognize that forceful "slum-improvement" programs may be counterproductive (World Bank 2009). Nevertheless, these state-centered perspectives on urbanization and development still tend to perceive squatter settlements as a transitory phenomenon that will be naturally dissolved over time, paying scant attention to what is currently happening in these urban spaces. From such a view, "education" is narrowly conceived to be provided through formal schooling; hence, lack of public/private schools are equated with the "lack of education" and institutions such as boxing camps would be completely outside the purview of "education." Yet, for the actually existing community of Fourth Estate, boxing camps can be said to provide the youth with critical skills, knowledge, and morality of mutual help that are every bit crucial to live well in the community.

References

Carrington, Bruce. 1986. "Social Mobility, Ethnicity and Sport." *British Journal of Sociology of Education*, 7 (1): 3–18.

Durkheim, Émile. 1965. *The Elementary Forms of the Religious Life*. New York: The Free Press.

Foucault, Michel. 2007. *Security, Territory, Population: Lectures at the Collège de France, 1977–1978*. Translated by Graham Burchell. New York: St Martin's Press.

Goffman, Erving. 1961. *Asylums: Essays on the Social Situation of Mental Patients and Other Inmates*. London: Aldine Publishing Company.

Ishioka, Tomonori. 2012a. *Rōkaru bokusā to Hinkon Sekai – Manira no BokushinguJim nimiru Shintai Bunka* [Local boxers and the World of Poverty – Body Culture in Boxing Gyms in Manila]. Kyoto: Sekai Shisō Sha.

———. 2012b. "Boxing, Poverty, Foreseeability – An Ethnographic Account of Local Boxers in Metro Manila, Philippines." *Asia Pacific Journal of Sport and Social Science* 1 (2–3): 143–55. https://doi.org/10.1080/21640599.2012.752666

Kitiarsa, Pattana. 2013. "Of Men and Monks: The Boxing-Buddhism Nexus and the Production of National Manhood in Contemporary Thailand." *New Mandala*. https://www.newmandala.org/pattana-kitiarsa-on-thai-boxing/

Kropotkin, Peter. 1902. *Mutual Aid: A Factor of Evolution*. New York: McClure Phillips & Co.

Matsuda, Motoji. 1998. *Urbanisation from Below: Creativity and Soft Resistance in the Everyday Life of Maragoli Migrants in Nirobi*. Kyoto: Kyoto University Press.

Roy, Ananya. 2005. "Urban Informality: Toward an Epistemology of Planning." *Journal of the American Planning Association*, 71(2): 147–58.

Singer, John N., and Reuben A. Buford May. 2011. "The Career Trajectory of a Black Male High School Basketball Player: A Social Reproduction Perspective." *International Review for the Sociology of Sport* 46 (3): 299–314.

Spaaij, R. 2009. "Sport as a Vehicle for Social Mobility and Regulation of Disadvantaged Urban Youth: Lessons from Rotterdam." *International Review for the Sociology of Sport* 44 (2): 247–64.

Torigoe, Hiroyuki. 2008. *"Sazae-san" teki Komyuniti no Hōsoku* [The Law of "Mrs. Sazae"-style Community]. Tokyo: NHK Publishing.

World Bank. 2009. *World Bank Report 2009: Reshaping Economic Geography*. Washington DC: The World Bank.

Part V
Critical Reflections and Prospects

15 Empirically Speaking

Life-Environmentalism, Environmental Justice, and Feminist Political Ecology

Daisaku Yamamoto, Sophia Ferrero and Keegan Kessler

In college-level social science courses in the United States, Environmental Justice and Political Ecology are commonly taught as concepts and bodies of literature that aim to critically analyze human–environment relationships and to inform socio-environmental actions and policies. Over the past few decades, numerous empirical studies have been conducted within these bodies of literature. Although they have somewhat distinct origins, both fields emerged in response to the perceived limitations of previously dominant theories of human–environment relationships, which often overlooked socioeconomic, cultural, and political dimensions.

In this chapter, our goal is to identify the distinctive features of life-environmentalism when viewed from the perspectives of Environmental Justice and Political Ecology. We also explore the potential areas of convergence among these approaches. Importantly, our focus is not solely on theoretical claims, but rather on how these claims are translated into concrete empirical studies within each approach. Within the realm of Political Ecology, we specifically concentrate on Feminist Political Ecology, taking into consideration the vastness of the Political Ecology literature and the noticeable similarities between Feminist Political Ecology and life-environmentalism.

In the following sections, we provide a brief overview of the origins of Environmental Justice and (Feminist) Political Ecology literatures, along with a summary of their key analytical emphases. We then examine how selected case studies of life-environmentalism, many of which are from this book, empirically address these issues, and compare their approaches to case studies informed by Environmental Justice (EJ) and Feminist Political Ecology (FPE). This chapter offers preliminary insights into potentially useful ways in which life-environmentalism may engage in further conversation with approaches in Anglophone academia.

Critical Perspectives on Human–Environment Relations in the Anglophone Social Sciences

Origins and Overview

Starting in the early 1900s, prominent American environmental activists like John Muir recognized the effects of human action on nature and called for equitable access to natural resources as well as environmental protection and animal rights (Taylor 2000). After WWII, the United States made significant technological advancements, fostering the belief that technology would provide human control over nature. Driven by the image of a "good life," resource consumption increased dramatically during this period (Kline 2011). Later, in the 1960s, Rachel Carson's book *Silent Spring* played a critical role in the first wave of the environmental movement, highlighting the consequences of unrestricted consumption and environmental destruction (Taylor 2000). American conservationists increased their efforts in response to elevated awareness of pollution and environmental concerns prompted by ecological disasters, such as the near extinction of the US national bird—the bald eagle (Kline 2011).

These early conservation movements, which later on became known as Traditional Environmentalism, aimed to relate environment quality with human values (Mollett 2017). However, these values predominantly reflected the views of white, middle-class activists who focused primarily on wilderness conservation, wildlife preservation, industrial pollution, and population control (Bullard 1990). This environmental elitism contributed to the devaluation and erasure of the voices of poor and non-white populations. Further, American cities have been recognized for their historical association with class and racial segregation, which has been physically manifested in practices of redlining and the siting of hazardous waste facilities across the country (Bullard 1990).

It was these reflections that led to the conception of the Environmental Justice (EJ) movement and framework (Bullard 1990). EJ, as a movement and model, began in the Southern United States from grassroots environmental movements and protests informed by the civil rights movement strategies, movement leaders, and activist networks that often focused on local issues such as soil and water contamination (Bullard 1990, 2021; Taylor 2014). With the publication of landmark reports that provided data that supported the connection between environmental hazards and race, these local movements spread nationally in opposition to redlining, zoning, park construction, and waste dumping in communities of color (United Church of Christ Commission for Racial Justice 1987; Bullard 1990; Bell and Carrick 2017). As a framework and movement, EJ has now grown to have international reach, and has been utilized to address local human–environmental issues around the world (Walker 2012).

The United States established the first Earth Day in 1970, followed by new environmental legislations such as the 1970 Clean Air Act and 1972 Clean Water Act. Alongside this, in response to the polarizing debates surrounding conversation policies between ecoscarcity and modernization theses, similar to life-environmentalism's response to natural environmentalism and modern technocentrism, the field of Political Ecology gained popularity as an alternative approach (Robbins 2012). It is important to note that the origins of Political Ecology are broader than is often recognized, with important agrarian works such as Blaikie (1985), Blaikie and Brookfield (1987), and Watts (1983a, 1983b) as paramount to the development of the field (Bridge et al. 2015). Nevertheless, the Euro-American environmental movements also played a significant role in the critical reflection of dominant understandings of nature–society relations, contributing to the field of Political Ecology (Bridge et al. 2015; Robbins 2012).

Meanwhile, a feminist turn within Political Ecology has prompted a reexamination of its assumptions to incorporate more critical inquiries into the systemic social inequities that shape people's experiences of and access to nature (Rocheleau et al. 1996; Elmhirst 2015). Feminist Political Ecology (FPE) considers gender as a critical variable in understanding human–environment relations that are inseparable from political power structures. Here, gender is not a predetermined social difference; rather, definitions and lived experiences of gender are constantly reproduced and rewritten through everyday interactions of politics and the environment (Mollett 2017). By examining the interactions between politics, economy, and nature with an emphasis on gender dynamics, feminist scholars come to understand how politics of gender disproportionately affect some people over others in terms of access to knowledge and knowledge production, access to resources, and participation in political processes (Elmhirst 2015). Further, feminist and FPE scholarship has embraced a more reflexive approach that acknowledges the role of community and academia in generating inclusive and critical research processes (Mollett 2017). For example, the notion of postcolonial intersectionality, as proposed by Mollett and Faria (2013), demands FPE scholarship a more explicit incorporation of race and racism to understand how patriarchy and gender relations are deeply rooted in postcolonial genealogies of racial power and "whiteness."

Key Characteristics and Points of Comparison

Just as the histories of EJ and FPE are entangled, the characteristics of these approaches exhibit both similarities and differences. In order to evaluate empirical studies of life-environmentalism, we have identified five key points that serve as a framework. These points pertain to how each approach understands and analyzes: (1) focal social groups, (2) underlying normative values, (3) power and politics, (4) key knowledge, and (5) bodily emotion. By examining these aspects, we can gain a deeper understanding of the shared traits and divergences between life-environmentalism and the Anglophone approaches.

Focal Social Groups

Both EJ and FPE place significant emphasis on addressing disproportionate environmental burdens faced by historically marginalized and vulnerable populations. EJ has traditionally focused on social groups defined by race, ethnicity, class, and income, as well as specific cases surrounding Indigenous peoples, differently abled people, future generations, and, most recently, incarcerated people (Walker 2012; Pellow 2016; Pellow 2021). Gender-based groups have been obviously a central concern within the FPE literature. Additionally, both literatures have increasingly recognized the importance of intersectionality, which means how identities such as race, class, gender, sexuality, citizenship, indigeneity, and physical ability are interconnected in a way that creates a complex and unique experience of oppression and discrimination for groups and individuals (Sze 2020, Mollett & Faria 2013; Perera-Mubarak 2013).

For example, in the wake of California's 2017 Thomas fire, undocumented Latinx and Indigenous communities faced disproportionate barriers in accessing adequate disaster aid and administrative support due to a combination of language barriers, racial discrimination, gender inequities in hazardous work environments, and fears of deportation (Méndez et al. 2020). This shows that various social subjectivities such as race/ethnicity and legal status compound with gender to generate distinct experiences of disaster impacts (Mollett & Faria 2013). In these ways, FPE and EJ provide frameworks to critically analyze how socially constructed differences contribute to uneven experiences and vulnerabilities among various groups.

Normative Values

While EJ is a dynamically evolving field of inquiries and practices (Agyeman et al. 2016), the pursuit of social justice continues to serve as its fundamental motivation for analyzing disproportionate experiences of environmental burdens. EJ primarily focuses on anthropocentric justice, seeking "justice for the people," especially those who are historically marginalized (Walker 2012, 5). However, there is also recognition within the literature of justice for nonhuman entities. In this perspective, everyone, including both humans and nonhumans, has an undeniable claim to "equal basic liberties" (Agyeman et al. 2016). Such liberties encompass the right to "be free from ecological destruction," "mutual respect," and public policy that does not include "discrimination or bias" (First National People of Color Environmental Leadership Summit 1991). In addition, there are also multiple conceptualizations of justice, including distributive, procedural, recognition, compensation, and retributive justice. Thus, while EJ now encompasses more nuanced understandings of justice, it remains grounded in a universal idea of justice.

In the FPE literature, there are fewer explicit conceptualizations or attempts of defining justice, though it does clearly recognize injustices experienced by marginalized people. The normative emphasis of FPE is on research and practice that "empowers and promotes social and ecological transformation for women and other marginalized groups" (Elmhirst 2018, 519). FPE critically examines the dominant political and economic structures of power that produce societal inequities by focusing on the voices, motivations, and perspectives of marginalized populations. It challenges hegemonic understandings of social and environmental processes, highlights the interests and experiences of marginalized people, evaluates the effectiveness of case-specific policy, and seeks to inspire political reforms or social movement (Bridge et al. 2015, 8; Rocheleau et al. 1996; Elmhirst 2015). In essence, one of the key goals of FPE is to empower marginalized groups in response to injustices they face.

Power and Politics

EJ and FPE both emphasize the analysis of power dynamics and how these dynamics shape access to and control over environmental resources as well as influence disproportionate exposure to environmental amenities or burdens (England 2003; Elmhirst 2015; Walker 2012). Both approaches highlight the role of state and corporate powers in producing and perpetuating socio-spatial inequities (Glassman 2010). To understand how global processes shape and are shaped by local environmental experiences, a growing number of studies in both fields adopt a multiscalar analysis of power, exploring social processes at the community level in relation to national and global political and economic dynamics (England 2003; Elmhirst 2015). In making these connections, researchers often employ concepts such as "environmental racism" (Nybo 2020; Taylor 2014), "heteronormativity" (Jauhola 2010), and "hegemonic masculinity" (Kimura and Katano 2014) to characterize the nature of these dynamics. FPE, perhaps more than EJ, penetrates deeper into the analysis of power dynamics at the household and individual levels through the use of embodied and subjective analyses (Davis & Hayes-Conroy 2018; Kimura and Katano 2014; Perera-Mubarak 2013). This approach seeks to uncover how power is manifested and experienced at the personal level, inscribing political meaning to local everyday experiences, by establishing connections to broader, extra-local processes (England 2003; Elmhirst 2015).

Knowledge

Feminist scholarship has long challenged the assumption of value-free objectivity and universality in science, and has advocated for the development of alternative ways of knowing based on everyday life, gendered experience, and explicit acknowledgement of values (Harding 1987). Feminist theory understands knowledge as subjective and fluid, recognizing the interconnectedness between experience and knowledge (Elmhirst 2015). Within this framework, Feminist Political Ecology (FPE) employs the concept of gendered environmental knowledge to capture the differences among individuals and groups of individuals in their environmental knowledge and practices, which are shaped by gendered hierarchies (Elmhirst 2015). More broadly, gendered environmental knowledge can be understood as a form of situational knowledge which recognizes the influence and embeddedness of history, culture, linguistics, and value on an individual's way of knowing (Momsen 2000). Thinking this way, FPE recognizes various other knowledge systems, such as Indigenous, Black, postcolonial, queer, embodied, and power knowledge systems (Elmhirst 2015; Momsen 2000; Longhurst 2001; Mollett 2017).

In the context of contemporary EJ research, there is also a growing recognition of the importance of local knowledge and the limits of traditional scientific approaches to understanding environmental risks, health effects, and community response (Porto and Finamore 2012). To address this issue, EJ uses the concept of "justice as recognition," which posits that injustice occurs when differences are not acknowledged or respected (Whyte 2017). For example, the failure to acknowledge differences in cultural and knowledge systems among Native Americans has had detrimental consequences, such as the inability to practice self-determined, traditional languages or subsistence methods (Whyte 2017; First National People of Color Environmental Leadership Summit 1991).

Emotion and the Body

Traditional Political Ecology scholarship generally does not explicitly challenge the conventional assumption that the mind and body are separate entities. However, influenced by feminist, queer, postcolonial, Black, Indigenous, and people of color ontologies, FPE challenges this way of thinking by centering lived experiences and emotions, humanizing and politicizing the body to give it intimate and subjective meaning. In FPE, the body is viewed as the smallest unit of scale onto which larger political and social processes are inscribed, and close attention to emotions and embodied experiences is justified as a way to challenge the "masculinist, rational, Western, Cateresian ontology of the separation between mind and body" (Zaragocin and Caretta 2021, 1505).

For example, the feminist geographical method of *cuerpo-territorio* ("body-territory"), used by Indigenous women in Latin America, maps how extractivist activity affects different parts of their body and describes corresponding emotions on paper (Zaragocin and Caretta 2021). They share their narratives and collective knowledge, grounded in embodied experiences and affectivity, to understand territorial conflicts and to initiate healing for their bodies. By recognizing the body and mind as interconnected, FPE grounds emotions and experiences to the body and understands how such subjective experiences are related to social structures (Elmhirst 2015; Hayes-Conroy and Hayes-Conroy 2013).

Meanwhile, while psychological impacts of pollution, environmental racism, and injustice were highlighted by Bullard (1990), questions regarding the body and emotions have only recently been explored in EJ scholarship. In recent "embodied EJ" research, influenced by Indigenous and feminist understandings of the body, researchers investigate injustice as violence upon the body and examine the impact of systems of power through emotions and feelings (Kojola and Pellow 2021). One of the pathways through which emotion and feeling have been discussed in EJ literature is by incorporating Native Cosmologies, which understand emotions, such as grief, anger, shame, and hopelessness, as being associated with environmental decline and confirming the function of systems of power (Norgaard and Reed 2017). Similarly, "Embodied Geographies of Environmental Justice" as coined by Gay-Antaki (2021), looks to seriously consider feminist perspectives of the body and self into EJ conversations of larger processes.

Assessing Life-Environmentalism Empirical Case Studies

Our interest is to explore the ways in which empirical studies of life-environmentalism address these five primary points of focus within the EJ and FPE perspectives. By doing so we identify distinct characteristics of life-environmentalism in comparison to the prevailing perspectives in Anglophone academia.

Focal Social Groups: Creatively on the Margins?

Being consistent to their theoretical emphasis on inequalities that shape exposure to environmental burdens and access and control over environmental resources, empirical case studies of EJ and FPE often focus on historically marginalized groups in society, defined by such sociodemographic factors as race, ethnicity, gender, indigeneity, and economic class. On the other hand, in case studies of life-environmentalism, while the issues of uneven environmental experiences are within their scope, the analysis of specific sociodemographic groups that are commonly studied in EJ and FPE seems less central.

For instance, Jauhola (2010) and Perera-Mubarak (2013), both FPE studies on post-tsunami reconstruction after the 2004 Indian Ocean Tsunami, focus on the unique experiences of women in the affected areas to understand how underlying political and social processes influence gendered understandings and experiences of recovery. These case studies seek to reveal uneven impacts of the disaster through an intersectional and gendered lens. On the other hand, the study of a tsunami-hit community in Tōhoku by Ueda and Torigoe (Chapter 9) focuses on articulating the community-level logic behind the response of Moune that arise from its historical experiences, rather than highlighting uneven impacts of the disaster within the community.

As an example of EJ research, Nybo (2020) investigates how commercial tourism development has threatened the unique cultural traditions and the livelihoods of the Gullah/Geechee people, descendants of enslaved West Africans, in the Sea Islands off the coast of the Southeastern United States. The study suggests a potential solution for the people to seek the protection of the Sea Island under the National Environmental Policy Act. On the other hand, Murata (Chapter 4) conducts a life-environmentalist analysis on coastal tourism development in Kamogawa, where the marina project triggered a local conflict among different stakeholders. Murata's focus is on how the community as a whole, despite the fundamental disagreements within, managed their everyday problems using the local social organization of *Teichi* and logic of *Kuchikiki*.

At glance, it may appear that life-environmentalism studies lack explicit focus on the environmental experiences of historically marginalized populations compared to the works of EJ and FPE. Indeed, if one examines the association between gender and perceptions of the post-tsunami reconstruction schemes in Moune, they may find that the voices of female residents are often sidelined. This is not to say, however, that life-environmentalism studies disregard social differences within communities all together. For example, Noda (Chapter 1) points to the critical voice of female residents in the *kabata* controversy; Murata (Chapter 4) examines contrasting views of fishermen, surfers, and local women regarding the marina project; and Yan (Chapter 7) highlights the role of the Bailishihui council made up of elderly residents.

Therefore, what seems to be characteristic of life-environmentalism case studies is not the lack of attention to marginalized social groups per se, but rather the aspects of these populations they focus on. These studies often emphasize the creative agency of those who are typically considered marginalized and burdened. Instead of focusing primarily on the disproportionate burdens these groups face and portraying them as victims, the case studies highlight their capacity to respond to tangible, everyday challenges in the community, even if their solutions may be partial and tentative.

Normative Values: Justice Concealed?

For those familiar with the literature of EJ and PE, the near-complete absence of the word "justice" in life-environmentalism studies may be noticeable. Indeed, some life-environmentalism studies intentionally refrain from using normative terms such as "justice" and "equity." For example, in his case study of a conflict over a proposed incinerator project in a Japanese community, Hirai (2018) cautions against relying solely on a seemingly "just" procedure in dealing with the NIMBY problem. He criticizes procedures that narrowly assume "just-ness" as a situation where everyone has an equal say in the decision-making process. He argues that a formularistic logic of equality risks overlooking a more genuine sense of fairness, which can only be grasped by understanding the unique local histories of communities and addressing "burden sharing" among residents.

In his study of the local conflict over the coastal development, Murata (Chapter 4) observes how a series of local conferences, where different stakeholders such as surfers and fishermen openly exchanged ideas based on the principles of democratic consensus formation, made the difference in their views more pronounced and positions solidified. Like Hirai, Murata also highlights the importance of paying close attention to the unique local history, which has formed community-specific knowledge and life organizations. Similarly, in the case of the Daegu's historical environmental preservation by Matsui (Chapter 13), what constitutes a "just" outcome would highly depend on one's perspective. On one hand, one can certainly see that the elimination of Japanese colonial legacies from the city landscape is the right thing to do, while on the other hand, as Matsui reveals, there are also understandable rationales, especially for the local residents, to keep some of the buildings from the colonial period.

Compared to these life-environmentalism case studies that seem reluctant to have a clear normative value guiding their studies, EJ studies are usually more explicit about their goal of pursuing justice and seem to have a clearer sense of what justice is. For instance, Zaferatos (2006) looks into the issue of a hazardous waste site in the reservation of the indigenous Swinomish people in Washington state (a situation somewhat similar to Hirai's case study), and ultimately argues that "tribal self-determination and tribal environmental management capacity building are two necessary and integral conditions before effective environmental justice can occur in Indian country" (907). Zaferatos certainly asserts the importance of understanding the local history of the Swinomish people, but there is little recognition of potential or actual intra-community divides and conflicts, and the focus is on the broader "injustice" in which the Swinomish people were excluded from critical decisions made about their land.

Even though life-environmentalism studies do not explicitly evoke such normative values as "justice," EJ advocates may argue that some life-environmentalism studies do effectively address social and environmental justice. For instance, Yamamuro's (Chapter 6) study of Tōkai Village can be interpreted as an analysis of villagers striving for procedural justice to incorporate more genuine voices of residents into the decision-making process regarding the incinerator project. While such an interpretation is possible, we suspect that life-environmentalists may consider notions such as "justice" more as elements of the logic of persuasion (*iibun*), rather than accurate descriptors of desirable outcomes that residents truly seek.

Power: Observed or Experienced

From their inception, power and politics have been integral to EJ and FPE. Recently, there has been an increasing recognition of the importance of cross-scalar analysis in various branches of geographic studies, including in the analysis of power relations (e.g., England 2003; Glassman 2010; Sheller 2018; Walker 2012). For example, in the aforementioned study of post-tsunami reconstruction processes, Jauhola (2010) examines how gender mainstreaming policies and advocacy at the national level reinforce heteronormative boundaries that restrict women's agency. She demonstrates how processes at larger geographical scales shape individual experiences and identity expression. Jauhola effectively calls for national policy reform and, more broadly, a critical examination of "normalized" social, political, and economic processes that contribute to disproportionate experiences and recovery capabilities.

In his seminal article, Torigoe (1989; Chapter 18) also emphasizes the importance of power within the life-environmentalism framework. However, in empirical case studies, there is limited evidence of full engagement of life-environmentalism with power and politics, particularly across different spatial scales. For instance, the two tsunami studies, Ueda and Torigoe (Chapter 9) and Kanebishi (Chapter 11), seem to accept problematic national policies as given conditions, with only brief critiques of current reconstruction policies, without subjecting them and associated politics to systematic analysis. Many other case studies in this book similarly depict nonlocal political actors and dynamics as a backdrop, with the possible exception of Kaneko (Chapter 10) who examines a series of national policies that undermined the resilience of rural villages to disasters.

From the perspectives of EJ and FPE, the analytical tendency of life-environmentalism appears to risk naturalizing ineffective governments and poor policies, treating them as if they are natural disasters and missing opportunities for broader policy critiques and reforms. In other words, life-environmentalism case studies may seem "uncritical." Nevertheless, life-environmentalists would likely respond to such criticisms by insisting on the importance of a subjective lens that focuses on how community members experience concrete manifestations of power dynamics in their daily lives, with the two-persons world of "I and thou" (Torigoe 1989; also Chapter 18, 269). Accordingly, empirical studies of life-environmentalism tend to prioritize the explanation of how and why certain policies or demands from powerful actors are perceived as "unreasonable" by the focal community. Rather than relying on abstract concepts such as "heteronormativity" (Jauhola 2010) or "hegemonic masculinity" (Kimura and Katano 2014) to identify and generalize structural problems from an external perspective, the emphasis is on understanding the subjective experiences and perspectives of the community members themselves on specific issues. While the difference between subjective and objective approaches to power is a matter of degree, it is still worth exploring how these different approaches contribute to more effective prescriptions for actions and policies in practice.

Knowledge: Using, Not Preserving

For students encountering life-environmentalism for the first time, the approach may appear to value traditional knowledge and practices, resembling scholarship that emphasizes the significance of "indigenous knowledge" and "traditional knowledge" in contemporary Euro-American sciences (e.g., Follett 2019; Gadgil et al. 1993; Lavigne et al. 2008; Mazzocchi 2020). In empirical studies of life-environmentalism, elements of indigenous and traditional knowledge and practices do receive attention, as seen, for example, in the kabata system in Noda (Chapter 3), the local knowledge about tsunamis in Ueda and Torigoe (Chapter 9), and the famine food in Kaneko (Chapter 10). However, it is also evident that empirical studies of life-environmentalism are not solely focused on traditional indigenous knowledge and practices.

First, key knowledge and practices focused in some of the life-environmentalism case studies are hardly "traditional." For example, in the life-environmentalism study of the post-disaster reconstruction processes after the Mount Unzen Fugen 1990 volcanic eruption in Nagasaki Prefecture, Nakamura (2018) documents how the affected local community has dealt with the new disaster prevention plan proposed by the national and prefectural governments. The study details how and why the community accepted the plan despite dissatisfaction over the spatial division created by the plan, and there is little reference to traditional or indigenous knowledge systems. This contrasts to studies such as Lavigne et al. (2008), conducted by French and Indonesian scholars on volcanic disasters and affected communities, which examines how Javanese people living near volcanoes perceive risks in culturally specific ways by drawing on examples such as local communities' cultural system and knowledge known as *korban* that regards the volcano as "two faced." As another example, Isogawa (Chapter 8) examines the use of the dry riverbed space known as *kasenjiki*, through the life-environmentalist lens. The dry riverbed is a recent creation of the statecraft in the late 1950s, and the residents' collective knowledge of commons governance is not inherently "traditional."

Second, life-environmentalism studies often focus on how local inhabitants utilize their locally specific knowledge in the forms of customary norms, rules, and practices to adapt their everyday life to new situations. For instance, in the study of the Chinese village, Yan (Chapter 7) describes how the local community reconstructed the local *Guandimiao* temple and its symbolic significance as an innovative way to deal with problems associated with urbanization. Similarly, Murata (Chapter 4) defends the *Teichi* fishing group and the *Kuchi-kiki* (mediation) practice not for their enduring presence, but for their practical adaptation in present-day situations. In other words, although life-environmentalism studies may often defend the value of locally rooted beliefs and practices, they rarely advocate for preserving them solely because of their historical longevity.

Third, the argument of Matsuda (2013) about everyday knowledge, or *bengichi* ("expedient knowledge") (Furukawa and Matsuda 2003) is noteworthy. According to Matsuda, expedient knowledge informs individuals about what knowledge (e.g., popular morality and living common sense) is appropriate to invoke in a given situation. In the case study of Harie, Noda (Chapter 3) describes how local women employed the analogy of fishing (okay to catch fish only for eating) and the sacredness of the *kabata* to persuade men not to build a tourist-*kabata*. In this case, what is important is not so much the culturally specific knowledge itself, but rather knowing when and how to utilize and invoke that knowledge. Yamamoto and Yamamoto (Chapter 5) also illustrate how the villagers possess knowledge that enables them to address the resident movement and engage with key actors against the golf resort development in a manner that is less conflicting and politicized.

Readers familiar with contemporary EJ and PE literature may still have questions about how life-environmentalism research accounts for potentially variegated knowledge based on sociodemographic attributes, such as gender, age, ethnicity, and economic class. For example, Ueda and Torigoe (Chapter 9) do not extensively explore potentially conflicting opinions and knowledge within households along gender or age lines regarding decisions to resettle near the coast, which are likely considered of significant importance in contemporary feminist research.

Embodied Emotions Everywhere

The roles of bodily feelings and affective dimensions have been a significant concern in the feminist scholarship (Davis and Hayes-Conroy 2018), and their importance is increasingly recognized in the EJ literature as well (Kojola and Pellow 2021; Norgaard and Reed 2017; Gay-Antaki 2021). For example, Kimura and Katano (2014) analyze farmers in Fukushima after the nuclear accident within a FPE framework. They highlight an episode of a food safety expert criticizing women for expressing concerns and confusion about radiation contamination in food. In this analysis, women's visceral fears and concerns were "censored as irrational and unscientific, while masculinized discourse propounded the heroism and patriotism of people who followed the official pronouncement" (114). In another example, Perera-Mubarak (2013) points out the drastic increase of "emotion work" following the tsunami disaster in Sri Lanka. This entailed not only the usual responsibilities of caring for children, the elderly, and the disabled but also the additional need to provide care for those who had lost their homes, suffered injuries, or were relocated due to the disaster. Perera-Mubarak underscores that women disproportionately shouldered this burden of emotion work.

The life-environmentalism framework also acknowledges the importance of emotion in socio-environmental issues (Torigoe 1989), although it is generally not explicitly framed in terms of gender. For example, Kanebishi (Chapter 11) focuses on the emotional desire of the people who have lost loved ones to "keep the ambiguous, ambiguous" as an important coping mechanism, and recognizes the phenomenon of apparitions as "real" for those who experienced it. In another case, Adachi (Chapter 12) examines the sentiments expressed by local residents who lament the loss of "natural elegance" in their touristified *Gujo* dance. Again, the subjective feelings of the residents play a pivotal role in their effort to revive their agency in the community festival. In his study of boxers in Manila, Ishioka (Chapter 14) reveals that one of the key motivations for these boxers to join in and remain in the boxing camps is the fear of "not knowing what happens next," or the loss of foreseeability, in their everyday life. These case studies highlight the significance of subjective emotions and shared feelings, which are rooted in the experiences of people in the communities.

The fact that life-environmentalism research centrally addresses the issues of embodied emotions without referencing the feminist literature or gender dimensions, as often seen in FPE research, may partly reflect the perspective of Eastern intellectual traditions. These traditions view the body and mind as inherently inseparable and tend to emphasize the affective dimension of intimate knowledge (McCarthy 2010). Although we should be careful not to overemphasize commonalities between life-environmentalism and feminist scholarship, there does seem to be potential areas of confluence and productive intellectual engagement.

Conclusions

This chapter attempts to capture the characteristics of life-environmentalism by examining its empirical studies through the lens of Environmental Justice (EJ) and Feminist Political Ecology (FPE). Specifically, we compare and contrast these studies in terms of how focal social groups, underlying normative values, power dynamics, key knowledges, and emotion are understood and analyzed. Through this focused comparison, we hope to have made the nature of life-environmentalism clearer for the Anglophone audience. As the number of empirical studies used is limited, we do not claim that our analysis is comprehensive. Nevertheless, we can draw a few tentative conclusions and prospects for future areas of intellectual engagement.

First, our assessment makes it clear that an emphasis on small geographic communities is a defining feature of life-environmentalism studies. Communities that life-environmentalism focus on are the domains where everyday life activities and relatively dense inter-human interactions occur, such as neighborhoods, villages, and small towns. In life-environmentalism studies, the community is not only the primary geographic focus but also the key unit where power dynamics are manifested, knowledge for securing everyday lives is formed, and shared emotions are mobilized for active agency. The significance of the community may be challenging to fully grasp for external observers because it often requires in-depth knowledge of local histories beyond what the residents are self-aware or can articulate clearly (e.g., Chapter 3). While the significance of small communities may decline with increased mobility of people and information, many empirical studies of life-environmentalism show that small geographic communities remain critical scales at which the well-being of individuals is secured and sustained.

Second, it is important to note that life-environmentalism's analytical emphasis on small communities does not mean that it views small communities as somehow unified, harmonious or "positive" entities. Empirical studies of life-environmentalism often shed light on the presence of divisions and conflicts within communities, and explore how these divisions and conflicts are handled. Consequently, while it is true that small communities may face structural exploitation or suppression by powerful entities such as the state and large corporations, life-environmentalism studies rarely advocate simply for granting them more autonomy and decision-making power. This nuance is intriguing considering that the "empowerment" of marginalized social groups is frequently emphasized in empirical studies of EJ and FPE.

Third, theoretically, life-environmentalism may appear human-centric and myopic, potentially overlooking broader issues of justice and long-term sustainability. Critics may argue that by prioritizing the well-being of community members, the local outcomes approved by life-environmentalism researchers, either explicitly or implicitly, could lead to globally suboptimal outcomes, such as the exploitation of other human and nonhuman communities. However, when examining concrete empirical case studies in life-environmentalism, there is little evidence that what are portrayed as "successful" outcomes at the community level do or are likely to result in global injustice and unsustainability (also see Chapter 16). While the absence of a clear blueprint for global justice and sustainability in life-environmentalism may be unsatisfying for some audiences, this "passive" orientation (or what can be described as a "weak theory" in Chapter 1) seems to be an inherent characteristic of life-environmentalism.

From the perspectives of EJ and FPE, even though the theoretical exposition of life-environmentalism may give the impression of being somewhat eccentric (Chapter 2), its empirical findings and insights are not incomprehensible, but rather quite understandable. In fact, the broad empirical orientations of life-environmentalism, such as the emphasis on adaptive local knowledge and the recognition of emotive drivers in community dynamics are compatible with recent EJ and FPE studies. This does not imply that one approach can or should overshadow the others; rather, we believe that it is essential to acknowledge and value the differences between the approaches while promoting ongoing conversation and exchanges of ideas and empirical insights.

References

Agyeman, Julian, David Schlosberg, Luke Craven, and Caitlin Matthews. 2016. "Trends and Directions in Environmental Justice: From Inequity to Everyday Life, Community, and Just Sustainabilities." *Annual Review of Environment and Resources* 41 (1): 321–40. https://doi.org/10.1146/annurev-environ-110615-090052

Bell, Derek, and Jayne Carrick. 2017. "Procedural Environmental Justice." In *The Routledge Handbook of Environmental Justice*, edited by Ryan Holifield, Jayajit Chakraborty, and Gordon Walker. 101–12. London: Routledge.

Blaikie, Piers. 1985. *The Political Economy of Soil Erosion in Developing Countries*. London: Routledge.

Blaikie, Piers, and Harold Brookfield. 1987. *Land Degradation and Society*. London: Methuen.

Bridge, Gavin, James P. McCarthy, and Tom A. Perreault. 2015. "Editor's Introduction." In *The Routledge Handbook of Political Ecology*, edited by Thomas Perreault, Gavin Bridge, and James P. McCarthy, 3–18. London: Routledge.

Bullard, Robert D. 1990. *Dumping in Dixie: Race, Class, and Environmental Quality*. Boulder, CO: Westview.

———. 2021. "Environmental Justice - Once a Footnote, Now a Headline," *Harvard Environmental Law Review* 45 (2): 243–48.

Davis, Sasha, and Jessica Hayes-Conroy. 2018. "Invisible Radiation Reveals Who We Are as People: Environmental Complexity, Gendered Risk, and Biopolitics after the Fukushima Nuclear Disaster." *Social & Cultural Geography* 19 (6): 720–40. https://doi.org/10.1080/14649365.2017.1304566

Elmhirst, Rebecca. 2015. "Feminist Political Ecology." In *The Routledge Handbook of Political Ecology*, edited by Thomas Perreault, Gavin Bridge, and James P. McCarthy, 519–30. London: Routledge.

———. 2018. "Feminist Political Ecologies–Situated Perspectives, Emerging Engagements." [Originally published as "Ecologías Políticas Feministas: Perspectivas Situadas y Abordajes Emergentes" in *Ecologia Politica* 54. https://www.ecologiapolitica.info/ecologias-politicas-feministas-perspectivas-situadas-y-abordajes-emergentes/] English translation available at: https://core.ac.uk/download/pdf/188257594.pdf

England, Kim. 2003. "Towards a Feminist Political Geography?" *Political Geography, Forum: Political Geography in Question* 22 (6): 611–6. https://doi.org/10.1016/S0962-6298(03)00065-9

First National People of Color Environmental Leadership Summit. 1991. "Principles of Environmental Justice." United Church of Christ Commission for Racial Justice.

Follett, Alec. 2019. "'A Life of Dignity, Joy and Good Relation': Water, Knowledge, and Environmental Justice in Rita Wong's Undercurrent." *Canadian Literature* 237: 47–63.

Furukawa, Akira, and Motoji Matsuda. 2003. *Kankō to Kankyō no Shakaigaku* [Sociology of Tourism and Environment]. Tokyo: Shinyōsha.

Gadgil, Madhav, Fikret Berkes, and Carl Folke. 1993. "Indigenous Knowledge for Biodiversity Conservation." *Ambio* 22 (2/3): 151–56.

Gay-Antaki, Miriam. 2021. "Grounding Climate Governance through Women's Stories in Oaxaca, Mexico." *Gender, Place and Culture: A Journal of Feminist Geography* 28 (9): 1234–57.

Glassman, Jim. 2010. "Critical Geography II: Articulating Race and Radical Politics." *Progress in Human Geography* 34(4): 506–12. https://doi.org/10.1177/0309132509351766

Harding, Sandra, ed. 1987. *Feminism and Methodology: Social Science Issues*. Bloomington: Indiana University Press.

Hayes-Conroy, Jessica, and Allison Hayes-Conroy. 2013. "Veggies and Visceralities: A Political Ecology of Food and Feeling." *Emotion, Space and Society* 6: 81–90.

Hirai, Yūsuke. 2018. "Meiwaku Shisetsu no Ukeire to Futan no Bunyū: Gomi Shori Shisetsu Yūchi o Kokoromita Shiga-Ken Hikone-Shi B shyūraku no Jirei Kara" [The Acceptance of a NIMBY facility and the Sharing of Burdens: The Case of Community B that Attempted to Attract a Waste Incinerator Facility in Hikone City, Shiga Prefecture]. In *Seikatsu Kankyō Shugi no Komyuniti Bunseki* [Community Analysis of Life-Environmentalism], edited by Hiroyuki Torigoe, Shigekazu Adachi, and Kiyoshi Kanebishi, 133–51. Kyoto: Minerva Publishing.

Jauhola, Marjaana. 2010. "Building Back Better? — Negotiating Normative Boundaries of Gender Mainstreaming and Post-Tsunami Reconstruction in Nanggroe Aceh Darussalam, Indonesia." *Review of International Studies* 36 (1): 29–50.

Kimura, Aya Hirata, and Yohei Katano. 2014. "Farming after the Fukushima Accident: A Feminist Political Ecology Analysis of Organic Agriculture." *Journal of Rural Studies* 34: 108–16.

Kline, Benjamin. 2011. *First Along the River: A Brief History of the US Environmental Movement*. Lanham: MD: Rowman & Littlefield.

Kojola, Erik, and David N. Pellow. 2021. "New Directions in Environmental Justice Studies: Examining the State and Violence." *Environmental Politics* 30 (1–2): 100–18.

Lavigne, Franck, Benjamin De Coster, Nancy Juvin, François Flohic, Jean-Christophe Gaillard, Pauline Texier, Julie Morin, and Junun Sartohadi. 2008. "People's Behaviour in the Face of Volcanic Hazards: Perspectives from Javanese Communities, Indonesia." *Journal of Volcanology and Geothermal Research* 172 (3–4): 273–87. https://doi.org/10.1016/j.jvolgeores.2007.12.013

Longhurst, Robyn. 2001. "Geography and Gender: Looking Back, Looking Forward." *Progress in Human Geography* 25(4): 641–48. https://doi.org/10.1191/030913201682688995

Matsuda, Motoji. 2013. "The Difficulties and Potentials of Anthropological Practice in a Globalized World." Translated by John Ertl and Sebastian Boehnnert. *Japanese Review of Cultural Anthropology* 14: 3–30.

Mazzocchi, Fulvio. 2020. "A Deeper Meaning of Sustainability: Insights from Indigenous Knowledge." *The Anthropocene Review* 7 (1): 77–93.

McCarthy, Erin. 2010. *Ethics Embodied: Rethinking Selfhood through Continental, Japanese, and Feminist Philosophies*. Lanham, MD: Lexington Books.

Méndez, Michael, Genevieve Flores-Haro, and Lucas Zucker. 2020. "The (In) Visible Victims of Disaster: Understanding the Vulnerability of Undocumented Latino/a and Indigenous Immigrants." *Geoforum* 116: 50–62.

Mollett, Sharlene. 2017. "Environmental Struggles are Feminist Struggles: Feminist Political Ecology as Development Critique." In *Feminist Spaces: Gender and Geography in a Global Context*, edited by Ann M. Oberhauser, Jennifer L. Fluri, Risa Whitson, and Sharlene Mollett, 155–87. London: Routledge.

Mollett, Sharlene, and Caroline Faria. 2013. "Messing with Gender in Feminist Political Ecology." *Geoforum* 45: 116–25. https://doi.org/10.1016/j.geoforum.2012.10.009

Momsen, Janet Henshall. 2000. "Gender Differences in Environmental Concern and Perception." *Journal of Geography* 99(2):47–56. https://doi.org/10.1080/00221340008978956

Nakamura, Kiyomi. 2018. "Hisaichi Niokeru Seikatsu Saiken – Nagasaki Ken Unzen Fugendake Funka Saigai Hisaichi no Jirei Kara" [Reconstruction of Life in Disaster-Affected Areas: A Case Study of the Unzen Fugen Volcanic Disaster in Nagasaki Prefecture]. In *Seikatsu Kankyō Shugi no Komyuniti Bunseki* [*Community Analysis of Life-Environmentalism*], edited by Hiroyuki Torigoe, Shigekazu Adachi, and Kiyoshi Kanebishi, 373–93. Kyoto: Minerva Publishing.

Norgaard, Kari Marie, and Ron Reed. 2017. "Emotional Impacts of Environmental Decline: What can Native Cosmologies Teach Sociology about Emotions and Environmental Justice?" *Theory and Society* 46 (6): 463–95.

Nybo, Paul N. 2020. "Environmental Justice and the Gullah Geechee: The National Environmental Policy Act's Potential in Protecting the Sea Islands Survey of South Carolina Law: Environmental Law." *South Carolina Law Review* 72 (4): 1039–66.

Pellow, David N. 2016. "Toward a Critical Environmental Justice Studies." *Du Bois Review: Social Science Research on Race* 13 (2): 221–36. https://doi.org/10.1017/s1742058x1600014x

———. 2021. "Struggles for Environmental Justice in US Prisons and Jails." *Antipode* 53 (1): 56–73. https://doi.org/10.1111/anti.12569

Perera-Mubarak, Kamakshi N. 2013. "Positive Responses, Uneven Experiences: Intersections of Gender, Ethnicity, and Location in Post-Tsunami Sri Lanka." *Gender, Place & Culture* 20 (5): 664–85. https://doi.org/10.1080/0966369X.2012.709828

Porto, Marcelo Firpo, and Renan Finamore. 2012. "Environmental Risk, Health and Justice: The Protagonism of Affected Populations in the Production of Knowledge." *Ciência & Saúde Coletiva* 17 (6): 1493–501.

Robbins, Paul. 2012. *Political Ecology: A Critical Introduction*. Second edition. Chichester, West Sussex: John Wiley & Sons.

Rocheleau, Dianne E., Barbara P. Thomas-Slayter, and Esther Wangari. 1996. *Feminist Political Ecology: Global Issues and Local Experiences*. London: Routledge.

Sheller, Mimi. 2018. *Mobility Justice: The Politics of Movement in an Age of Extremes*. London: Verso.

Sze, Julie. 2020. "Introduction: Environmental Justice at the Crossroads of Danger and Freedom." In *Environmental Justice in a Moment of Danger*, 1–24. Oakland, CA: University of California Press.

Taylor, Dorceta E. 2000. "The Rise of the Environmental Justice Paradigm: Injustice Framing and the Social Construction of Environmental Discourses." *American Behavioral Scientist* 43 (4): 508–80. https://doi.org/10.1177/0002764200043004003

———. 2014. "Toxic Exposure: Landmark Cases in the South and the Rise of Environmental Justice Activism." In *Toxic Communities: Environmental Racism, Industrial Pollution, and Residential Mobility.* New York: NYU Press.

Torigoe, Hiroyuki, ed. 1989. *Kankyō Mondai no Shakai Riron: Seikatsu Kankyō Shugi no Tachiba Kara* [Social Theory of Environmental Problems: From the Standpoint of Life-environmentalism]. Tokyo: Ochanomizu Shobō.

United Church of Christ Commission for Racial Justice. 1987. "Toxic Wastes and Race in the United States: A national Report on the Racial and Socio-economic Characteristics of Communities with Hazardous Waste Sites." New York: United Church of Christ.

Walker, Gordon. 2012. *Environmental Justice: Concepts, Evidence and Politics.* London: Routledge. doi:10.4324/9780203610671

Watts, Michael J. 1983a. "On the Poverty of Theory: Natural Hazards Research in Context." In *Interpretations of Calamity: From the Viewpoint of Human Ecology*, edited by Kenneth Hewitt, 231–62. Boston: Allen & Unwin.

———. 1983b. *Silent Violence: Food, Famine, and Peasantry in Northern Nigeria.* Berkeley: University of California Press.

Whyte, Kyle. 2017. "The Recognition Paradigm of Environmental Injustice." In *The Routledge Handbook of Environmental Justice*, edited by Ryan Holifield, Jayajit Chakraborty and Gordon Walker. 113–23. London: Routledge.

Zaferatos, Nicholas C. 2006. "Environmental Justice in Indian Country: Dumpsite Remediation on the Swinomish Indian Reservation." *Environmental Management* 38 (6): 896–909. https://doi.org/10.1007/s00267-004-0103-0

Zaragocin, Sofia, and Martina Angela Caretta. 2021. "Cuerpo-Territorio: A Decolonial Feminist Geographical Method for the Study of Embodiment." *Annals of the American Association of Geographers* 111 (5): 1503–18. https://doi.org/10.1080/24694452.2020.1812370

16 Life-Environmentalism, Critiques, and Prospects

Focusing on the Experientialist Approach

Yasushi Arakawa

The concept of life-environmentalism originated from a research project conducted in the early 1980s in Chinai Village, a small settlement on the north coast of Lake Biwa. The study was a collaborate effort among a group of researchers with the goal of gaining insights to address the environmental problems of Japan's largest lake, particularly water pollution, which was a significant concern at the time. Rural village research in Japan has a rich history that can be traced back to the pre-WWII era, and it has yielded a substantial body of monographic studies, including comprehensive all-household surveys conducted in various parts of the country. The Japanese Association for Rural Studies, established in 1952, has played a key role as an academic organization in rural research. Since its inception, it has fostered cross-disciplinary studies by bringing together researchers from multiple fields. Following this cross-disciplinary tradition, the Lake Biwa research team, led by Hiroyuki Torigoe, consisted of researchers from diverse disciplines such as sociology, anthropology, folklore studies, and human geography.

The concept of life-environmentalism has been shaped by the multidisciplinary research team, and as a result, different members of the team have offered varying interpretations and emphases on the concept. In this chapter I focus on the works of Torigoe because they provide important insights into the origins of life-environmentalism within the framework of knowledge that was largely developed in Japan. Torigoe's approach builds on his intellectual background in the Japanese folklore studies established by Kunio Yanagita (1875–1962). Torigoe further refined his own methodology under the guidance of Kizaemon Aruga (1897–1979), a sociologist who also studied with Yanagita.

Japanese folklore studies, often referred to as "Yanagita folklore studies" in honor of its founder, focused on comprehending the everyday lives of common people (常民: *jōmin*) through meticulous fieldwork and historical sources. Its aim is to elucidate the cultural characteristics of ordinary Japanese people, rather than those of the elites (such as aristocrats and former samurai class) that had been frequently and erroneously portrayed as representative of Japanese culture.

In align with this tradition, Torigoe ventured into rural areas of the country, but his goal was to identify insights to address environmental problems by examining the ways in which people live their everyday lives. In essence, life-environmentalism can be viewed as an application of the method strongly influenced by the Japanese folklore studies to environmental issues.

To articulate the distinctive features of life-environmentalism, specifically as interpreted by Torigoe, this chapter begins by addressing two major criticisms levelled against the life-environmentalism research. I will then respond to these criticisms, drawing particularly on *keiken-ron* ("experience-approach"), or what can be tentatively translated as the "experientialist approach," which finds its root in the Japanese folklore studies. Through this discussion, this chapter explores the potential of life-environmentalism in addressing present-day environmental challenges.

Two Main Criticisms of Life-Environmentalism

Life-environmentalism has faced various criticisms since its original introduction in *Mizu to Hito no Kankyōshi* (Environmental History of Water and People) (1984). Here I address two important criticisms, which help to highlight the characteristics of life-environmentalism. The first criticism revolves around the researcher's supposed standpoint. Torigoe (1997, 11) asserts that life-environmentalists "adopts the position of the (everyday) lifeways of the people who live in the society in question," and advocates for a strong focus on inhabitants' "(everyday) life" to find solutions to existing problems. Regarding this claim, critics like Koichi Hasegawa (1996) contends that life-environmentalists would have to "introduce some kind of normative value standard into the field and interpret 'this behavior or lifestyle is in line with life-environmentalism' based on the standard" (131; author's translation). In other words, while life-environmentalists claim to adopt the standpoint of inhabitants (or their everyday life), they implicitly select certain inhabitants whose actions align with the investigator's normative value (such as those enthusiastic about environmental conservation or about recycling programs) and represent them as representative of all inhabitants. Hasegawa contends that in doing so, life-environmentalists obscure the normative values upon which they stand. This criticism concludes that if that is indeed the case, this criticism concludes, life-environmentalists must explicitly clarify which claim of the inhabitants they align themselves with.

The second criticism posits that life-environmentalism lacks a critical stance toward power, which may result in ineffective environmental policies. Kōichirō Miura (1995), for instance, points out that "life-environmentalism desperately lacks defense against intervention by external power," and that it overly emphasizes the community while holding an uncritical faith in village community norms. Miura criticizes this approach a kind of epoché, or the suspension of judgment, which "will only justify grassroots fascism" (Miura 1995, 473, 483; author's translation).

I suspect that both criticisms are related to the fact that life-environmentalism sets *keiken-ron*, or the experientialist approach, as its methodological foundation. Torigoe explains that "experience" in this context means "the accumulation of the remembered time from the past for individuals or groups" and that life-environmentalist analysis focuses on "experience" that serves as the basis for selecting a particular "action" among various options, rather than analyzing the observed actions themselves (Torigoe 1997, 20–23; author's translation). Experience, by its nature, is the accumulated totality of how individuals and groups responded to various circumstances over time. Therefore, for life-environmentalist analysis, understanding what inhabitants have done in the past, and how they have done it, becomes crucial, rather than solely focusing on their current actions. However, if this is the sole focus, policy measures would likely merely endorse what inhabitants have already done, resulting in the confirmation of the status quo and ineffective policies, which aligns with the point raised in the second criticism. The fact that policy arguments informed by life-environmentalism are sometimes mocked as a "romantic regression to the past" (Hasegawa 1996, 130) is closely linked to the methodological characteristics of life-environmentalism that use experience as the foundational unit of analysis.

Experientialist Approach and Its Implications

Both of the criticisms described above raise objections to the methodological core of life-environmentalism, which involves standing on or by the position of the lifeways of the people living in a particular community, and using their experience as the basis for analysis. Thus, in order to explore the future prospects of life-environmentalism, it is necessary to defend the experientialist approach itself. To do so, let us dig deeper into the essence of the experientialist approach and identify more precisely where these criticisms are directed.

When life-environmentalists aim to contribute resolving an environmental problem in a local community through research, they would visit the community and seek to study the lived experience of the inhabitants. However, they are likely to encounter a multitude of diverse ways of living, carried out by unique individuals who differ in appearance, thoughts, and actions. Life-environmentalists are expected to analyze the inhabitants' "experience" by taking the position of the lifeways of the people in that society. But is it even possible to adopt such a position when each inhabitant is a unique individual?

This is where the methodology developed in Japanese folklore studies becomes valuable. Specifically, what characterizes Kunio Yanagita's folklore research is the perspective of viewing the subject not through intellectual analysis, but through *kokoro*, or "the mind." According to Torigoe, Yanagita folklore "adopts the method of interpreting the subject by consistently unifying the minds of the self and the subject as one" (Torigoe 2002, 181; author's translation). The notion of *kokoro* is quite similar to the concept of *mentalités* in the French Annales school. However, there are significant differences in the underlying logic between *mentalités*, which are grasped through the unique method of historical understanding refined by the Annales, and *kokoro*, which has its intellectual origins in the poet theory of medieval Japan and continues through the works of Yanagita (Torigoe 2002).

Life-environmentalism researchers aspire to understand why people in a community facing an environmental problem live the way they do today. To find an answer, it is necessary to study the history of how those people have lived in the community. In Yanagita folklore studies, the scope of history does not have to extend to the ancient past; rather, it emphasizes the importance of studying a relatively recent past that directly influences the present. As Yanagita states (1935 [1990]), we must closely examine "how our parents and their parents lived, just 100 or 200 years ago" (14; author's translation) because their consciousness likely still resonates in parts of our own consciousness and influences our thoughts and actions.

Accordingly, when conducting fieldwork following the life-environmentalist approach, seemingly efficient survey methods like collecting data through questionnaires focused solely on the focal problem will not be the central focus of the study, although they may serve as supplementary sources of information. Instead, the researcher extends the study to various inter-human and human-environment relationships that may seem unrelated to the focal problem at first (such as customs related to water usage and local shrine festivals). Through this approach, they strive to grasp the broad, diachronic context of the everyday life of the inhabitants. Only then is it possible to understand the reasons and circumstances behind why the people's lives have evolved into their current state. This type of analysis, as Torigoe (1997) describes, "descends to the experience at the root of the [observed] action and from which an analysis starts" (22; author's translation).

In this way, life-environmentalism research requires researchers to closely observe various inter-human and human–environment relationships through repeated fieldwork in the study area. During this process, researchers must refrain from judging what the inhabitants do based on existing knowledge and value systems from a third person's point of view. Instead, they must try to align their mind (*kokoro*) with that of the inhabitants and explore the circumstances that have shaped their current ways of life. This is what is meant by studying from "the position of the (everyday) lifeways of the people who live in the society in question."

It is important to note that the current way of life in the study area is not solely the result of rational decisions made by community members over time. Factors such as local politicians and municipal governments, conflicts with neighboring communities, influential visitors from outside, or natural disasters can all significantly alter the ways of life in the community. Inhabitants, whether individually or in organized groups, have been responding and adapting to various externally induced changes in order to sustain their livelihoods. In essence, those with life-environmentalist perspectives closely examine how people in the community arrive at their observed actions, understanding that they live within a web of diverse relationships and with an emotional sensitivity toward others. By doing so, life-environmentalism seeks to grasp the experiences of each individual and group living in the community. Life-environmentalists do not adhere to a narrow scientific approach that posits a rational individual and behavior, and then measures the deviation from actual behavior. Instead, they believe that by avoiding such a narrow scientific approach, they can arrive at a better understanding the reality of the everyday lives of people in the community.

Life-Environmentalists' Response to the Criticisms

Based on the premise that life-environmentalism analyzes the everyday life of the people who live in the complex web of relations from their point of view, let us revisit the two criticisms mentioned above. The first point pertains to the standpoint of researchers, charging that life-environmentalists selectively choose inhabitants whose actions align with their own normative value and represent them as representative of "the inhabitants" while concealing their own position. However, in reality, life-environmentalists analyze the everyday life of inhabitants by descending to their experience underlying their actions, and the formation of *iibun* (logic of justification and persuasion) within resident groups actively dealing with specific concrete problems in the local community. In other words, there are no inhabitants whose "behavior or lifestyle is in line with life-environmentalism" (Hasegawa 1996, 131) from the outset. What can exist is a life-environmentalist analysis of inhabitants who deal with specific concrete problems from their own standpoints.

Accordingly, life-environmentalist analysis cannot exist without the analysis of micro-level, specific relationships in the area in question. Only through such observations is it possible to comprehend the conditions under which people arrive at their actions. For this reason, "ordinary everyday life" of the inhabitants must be an integral part of life-environmentalist analysis. For those who envision sites of environmental problems being place of intense political conflicts and severe disruptions to daily life, the depiction of "ordinary everyday life" in life-environmentalist account may not seem to be reflective of reality. However, from the perspective of life-environmentalism, even in the presence of intense conflicts among inhabitants, researchers do not solely focus on the conflict itself. Instead, they recognize the importance of understanding the everyday experiences of individuals that underlie the conflict. It is precisely for this reason that "theories" (in a broader sense of logical explanations; see also Chapter 18) based on "ordinary everyday life" offer practical value that abstract theories lack.

At this point, it is necessary to address how life-environmentalism understands the notion of "utility" of policies derived from life-environmentalism. According to Torigoe and Kada (1984),

> First and foremost, "theory" is not a "real" entity ground in a collection of facts; it is merely an analytical framework used to interpret social phenomena. In essence, it is a "fiction." There are instances when the theory as a fiction proves to be useful. In such cases, society determines the

theory as having a degree of reality. It is worth nothing, by the way, that the criterion for judging the reality of a theory is not based on data; rather, it is based on the sense of everyday life.

(Torigoe and Kada 1984, 338; author's translation)

For them, the utility of a theory, including life-environmentalism, is ultimately determined by the "sense of everyday life." If one wishes to be true to this criterion, they must construct a "theory" that is based on "facts" perceived as realistic by individuals living their everyday lives in the community. Only when a "theory" faithfully reflects the "facts" that resonate with the sense of everyday life can it be truly useful. This implies that the utility of a policy derived from a "theory" must also be determined by the people who are deeply invested in solving the community's problems.

The second criticism revolves around the lack of critical analysis of power in life-environmentalism research. It is true that when we explore the everyday lives of people, closely examining concrete social reality instead of relying on simplistic abstraction, we often encounter the existence of power dynamics within concrete relationships among human actors, as well as between human and nonhuman actors. Power dynamics often operate overtly or covertly beneath the surface of what may appear to be idyllic daily life, manifesting, for example, as asymmetric power relationships between households, the influence of "village norms," the economic dominance of companies or local elites responsible for pollution, and the influence of municipal or national governments over local communities. In fact, the functioning of asymmetric power relationships is actually a normal state of affairs in communities grappling with various environmental issues.

Life-environmentalists undoubtedly encounter these power relationships when they attempt to depict the everyday lives of inhabitants. However, they do not typically engage in direct analysis of powerful actors such as governments and companies, nor do they evaluate the effects of power from a bird's-eye view. Instead, they discern the subtle workings of power among the people at the site of environmental problems who make judgments and decisions based on the existence of powerful actors. During these instances, power does not exist in the abstract but operates tangibly behind people's actions. Life-environmentalists possess a sensitivity to the concrete manifestation of power in such contexts.

Because of these characteristics, life-environmentalists only criticize the inhabitants themselves in limited situations where their actions clearly pose a threat to their own everyday lives (e.g., clear-cutting forests in their water source area). It goes without saying that such self-destructive actions are not uncommon in environmental problem sites, and it is precisely because of this that life-environmentalist analyses, grounded in the experientialist perspective, can provide critical insights and unique practical value. In essence, life-environmentalism, which prioritizes the analysis of ordinary everyday lives, may sometimes be subject to ridicule as "a theory of an idyllic and peaceful world."

However, it is also an approach that remains deeply connected to the social reality of the community under scrutiny, making it highly practical in nature.

Life-Environmentalism's Continuing Relevance

More than 30 years have passed since the birth of life-environmentalism, and during this time, the landscape of environmental issues has undergone significant changes. Notably, neoliberal ideologies have deeply permeated various policy domains, and regional policies that were once dominated by local governments now encourage much greater civic participation. Additionally, the concept of "sustainable development" has become fully integrated into national and regional policy discourse. The popularity of "sustainable development" surged in the 1980s, particularly when it became a central concept in *Our Common Future* (1987). Despite receiving various criticisms, it has been embraced as a viable concept by global institutions, policymakers, and academics.

When "sustainable development" is put into practice as a policy, it typically involves the formulation of plans by national governments based on scientific knowledge, evidence, and international consensus. Businesses and citizens are then expected to comply with or collaborate on implementing these plans. Consequently, the definition of "sustainable development" and its concrete practices are primarily determined by scientists with global and national perspectives, with final decision-making power residing in national governments or international agreements. As a result, "sustainable development" is often implemented in a top-down manner, with citizens expected to follow decisions made "from above."

Life-environmentalism is often critical of this type of top-down policy approaches. However, at first glance, both life-environmentalism and the principles of "sustainable development" may appear to endorse similar goals and practices such as sustainable resource management, consideration of the well-being of future generations, and the coexistence of human and nonhuman beings. One may wonder therefore if life-environmentalism has lost its novelty, given these similarities. However, as discussed above, the value of life-environmentalism lies primarily at the methodological level, grounded on the experientialist approach, rather than solely in the (desirable) actions and practices themselves. The following simple illustration shows how the same action by local inhabitants may be interpreted differently by life-environmentalists and the proponents of "sustainable development."

A mountain village in Hyogo Prefecture, like many mountain villages in Japan, had a thriving sericulture industry from the Meiji period to the early Showa period. However, due to rising raw material and labor costs, and increased competition from international markets, it became difficult to make a living from raw silk production. As a result, partly influenced by the government's afforestation policy at the time, the local residents began planting cedar and cypress trees in the mountains. Today, these trees cover the areas near the village all the way to the mountaintop. However, despite the arrival of the harvest season, the trees are rarely harvested. This is primarily because logging is not economically viable due to low timber prices, and this trend is observed nationwide.

However, in this particular village, plans have emerged to cut down the coniferous trees, such as cedar and cypress, and replace them with broadleaf trees like oak and sawtooth oak. From the perspective of "sustainable development," this initiative is likely to receive positive evaluations. The potential natural vegetation in the area consists mainly of broadleaf trees, and the restoration of such an environment would likely be more sustainable and stable, requiring less human intervention in the long term. Additionally, multilayered forests comprising various broadleaf tree species contribute to biodiversity and align better with national environmental policies compared to single-layered forests dominated by a single tree species. Furthermore, the plans were initiated by the local community, and this apparent bottom-up approach is in line with recent trends in sustainable development policies. Therefore, if the afforestation plan for this area is implemented, the initiative may be viewed as a progressive model of "sustainable development" practices.

Yet, when I talked to people in this area about their plan to replant broadleaf trees, it became evident that the inhabitants do not necessarily understand this plan to be the act of environmental conservation, as proponents of "sustainable development" may assume. The rationale of the people of the village can be roughly summarized as follows:

> Our lives are constantly threatened by deer and bears. In the past, deer and bears were afraid of humans and rarely ventured near houses because they were able to find sustenance in the mountains. While afforestation with broadleaf trees is not economically viable for us, we believe it is the best way to ensure our current livelihoods.

In other words, for the local people, the plan to replant trees with broadleaf species is aimed at providing food for the wild deer and bears in the mountains, thereby safeguarding their own well-being. Therefore, if outsiders were to visit the village and enthusiastically declare, "It's wonderful that you are actively implementing 'sustainable development' practices," the inhabitants would likely be perplexed or respond with wry smiles, realizing that their motivations differ (Arakawa 2009).

Various initiatives and activities that seem consistent with the perspective of "sustainable development" are being carried out across the country. However, a closer examination reveals that these efforts do not always conform to the idea of "sustainable development" as deducted from the analysis of global environmental problems. While the individuals involved in these activities, such as afforestation, may be clearly aware of the concept of "sustainable development" and may take advantage of associated opportunities, such as subsidies for tree replanting, it is important to recognize that their actions are not driven by the idea of "sustainable development" itself but are instead based on judgments grounded in their lived experiences in the area.

It is through these "discoveries" of the lived experience of the inhabitants that life-environmentalists construct a "theory" that resonates with the "sense of everyday life" and contemplate how to arrive at policies and actions to solve the problem at hand.

Prospects for Life-Environmentalism

Matsumura (2007 also see Chapter 17) suggests that the perceived uniqueness of life-environmentalism, initially distinguished from natural environmentalism and modern technocentrism, may no longer be as novel, as a resident-centered approach has become the new norm in most environmental policies. Wakita (2005) also advocates for more dialogue and collaboration with other approaches and fields of inquiry, rather than competing with them. Torigoe, who has been at the forefront of life-environmentalism, is fully aware of these changes in environmental discourses, but cautions that achieving such dialogue and collaboration are not easy. According to Torigoe:

> It would be overly simplistic to suggest that all we need to do is to create a policy approach based on actual everyday life. I believe we need a different kind of policy approach when dealing with environmental problems in our country. It should be a policy approach that can be considered more real than reality itself, breathing life to the system of everyday logic generated from deep local history. That is, I believe it is necessary to formulate policy approaches based on a more familiar, lived world that often includes aspects that are not rationale, rather than solely relying on the current rational way of life.... To develop concrete policies, it is absolutely essential to understand how the "actors responsible for nature conservation" (primarily local inhabitants) perceive nature and nature conservation. A desirable path is to generate viable policies by following the ideas of these actors while also listening to the opinions of scientists and the government.
>
> (Torigoe 2001, 15, 20)

Even today, in the realm of policymaking, it is still common for policymakers to involve residents as a convenient justification, or alibi, under the guise of public participation (Adachi 2001). Torigoe's statement is clearly a caution against such superficial "public participation" and "public-private partnerships," but it carries further implications. When life-environmentalism proposes an environmental policy based on the experientialist approach, it implies that the policy should be informed by and selected from the depths of people's experiences. As mentioned earlier, from a life-environmentalist perspective, the usefulness of a policy derived from a "theory" ultimately lies in the hands of the individuals who earnestly seek to solve the problems within their community. However, this does not simply mean that residents should have the power to approve or reject a particular policy (e.g., through voting). Rather, inhabitants, who are entrusted with scrutinizing the policy, are expected to make policy decisions based on deep reflections of their experiences, intertwining the past and the future. This is why Torigoe emphasizes the aforementioned caveat; it is not merely a matter of involving residents early in the decision-making process or giving them the final say. Instead, from a life-environmentalist perspective, a critical examination through the "experience" of inhabitants is essential to devise a policy theory from "a more familiar, lived world."

The fact that life-environmentalism still maintains its unique position relative to various other approaches can be observed in its stance on environmental policy, and the relevance and applicability of life-environmentalism to contemporary environmental problems should be understood as an extension of this distinct orientation.

References

Adachi, Shigekazu. 2001. "Kōkyō Jigyō o Meguru Taiwa no Mekanizumu – Nagara Gawa Kakozeki Mondai o Jirei Toshite" [Mechanism of Dialogue on Public Works: A Case Study of the Nagara River Estuary Barrage Problem]. In *Higai to Kaiketsu Katei: Koza Kankyō Shakaigaku Dai 2 Kan* [Damage and Resolution Process: Seminar Environmental Sociology Vol. 2], edited by Harutoshi Funabashi, 145–76. Tokyo: Yūhikaku.

Arakawa, Yasushi. 2009. "Seikatsu Kankyō Shugi niokeru 'Gendaisei'–'Jizoku Kanō na Shakai' Ron tono Kankei o Chūshin ni" [Today's Implications of Life-Environmentalism: Focusing on Sustainable Society]. *Gendai Shakaigaku Riron Kenkyū* [Journal of Studies in Contemporary Sociological Theory], 3: 28–37.

Hasegawa, Koichi. 1996. "Shohyō Kada Yukiko Cho–Seikatsu Sekai no Kankyōgaku" [Book Review: Kada, Yukiko "Environmental Studies in the Living World"]. *Soshioroji* [Sociology], 41 (2): 128–31.

Matsumura, Masaharu. 2007. "'Seikatsu Kankyō Shugi' Ikōno Kankyō Shakaigaku notameni" [For Environmental Sociology After "Life-Environmentalism]. In *2003–2006 KAKEN Grant-In-Aid For Scientific Research (B) Report: Comprehensive Study on Environmental Problems in Japan and Asia-Pacific Region with The Historical Study on the Theories and Research Methodologies of Japanese Environmental Sociology*, edited by Yōsuke Hoashi. 273–88.

Miura, Kōkichirō. 1995. "Kankyō no Teigi to Kihanka no Chikara–Naraken no Shokuniku Ryūtsū Sentā Kensetsu Mondai to Kankyō Hyōshō no Seisei" [Normative Power and the Definition of the Environment: Meat Distribution Center Construction Problems in Nara Prefecture and the Creation of Environmental Representation]. *Shakaigaku Hyōron* [Sociological Review], 45: 469–85.

Torigoe, Hiroyuki. 1997. *Kankyō Shakaigaku no Riron to Jissen: Seikatsu Kankyō Shugi no Tachiba kara* [Theory and Practice of Environmental Sociology: Perspectives of Life Environmentalism]. Tokyo: Yūhikaku.

———. 2001. "Ningen ni Totte no Shizen–Shizen Hogoron no Saikentō" [Nature for Humans: Reexamination of Nature Conservation Theory]. In *Shizen Kankyō to Kankyō Bunka: Koza Kankyō Shakaigaku Dai 3 Kan* [Natural Environment and Environmental Culture: Seminar Environmental Sociology Vol. 3], edited by Hiroyuki Torigoe, 1–23. Tokyo: Yūhikaku.

———. 2002. *Yanagita Minzokugaku no Firosofi* [English title: *Yanagida Kunio and Japanese Folklore*]. Tokyo: University of Tokyo Press.

Torigoe, Hiroyuki, and Yukiko Kada, eds. 1984. *Mizu to Hito no Kankyōshi: Biwako Hōkoku Sho* [Environmental History of Water and People: Lake Biwa Report]. Tokyo: Ochanomizu Shobō.

Wakita, Kenichi. 2005. "Biwako Nōgyō Dakusui Mondai to Ryūiki Kanri 'Kaisōka sareta Ryūiki Kanri' to Kōkyōken toshite no Ryūiki no Sōshutsu" [Lake Biwa's Agricultural Muddy Water Problems and Watershed Management: Hierarchical Watershed Management and Creation of Watersheds as Public Spheres]. *Annual Reports of the Tohoku Sociological Society* 34: 77–97.

Yanagita, Kunio. 1935. *Kyōdo Seikatsu no Kenkyūhō* [Methods of Research on Local Life], Tokyo: Tōkō Shoin.

17 The Future of Life-Environmentalism
A Sympathetic Critique

Masaharu Matsumura

Life-environmentalism is a key approach that is well known today within Japanese environmental studies, especially environmental sociology. When studying environmental issues in Japan, one cannot ignore this body of literature and its surrounding debates. Life-environmentalism originated in the 1980s, when a relatively cohesive theory and methods arose from intensive fieldwork in the Lake Biwa region, and has informed a large number of empirical case studies since then (Kada 1995; Torigoe 1997; Torigoe and Kada 1984; Torigoe et al. 2018). Setting its contributions aside, life-environmentalism has also received various criticisms from within Japanese academia. Many of the criticisms did not lead to productive debates, but some raised important points, which this chapter explores.

In addition, as decades have passed since the birth of life-environmentalism, some of its insights are no longer as novel as when the approach was first introduced. There have also been some changes in its key propositions over time. Furthermore, there is a growing interest in global-scale environmental problems as seen in the increasing awareness that we are in the Anthropocene, a new geological epoch in which humankind is exerting significant influence on the Earth's climate and ecosystems. Many government and nongovernment actors are also taking part in the United Nations' 2030 Agenda for Sustainable Development, including the 17 Sustainable Development Goals (SDGs). In times like this, does life-environmentalism still offer relevant insights? If so, what might they be?

This chapter discusses key points of debate surrounding life-environmentalism, explores its relationship with modern environmental sociological theories and environmental thought, and seeks to deepen our understanding of this approach. By doing so, I hope to identify what life-environmentalism may still have to offer us today.

Is Life-Environmentalism Outdated?

The Emergence of Life-Environmentalism: Beyond the Human-Nature Binary

Life-environmentalism is a research approach proposed by a group of Japanese sociologists and anthropologists who conducted intensive field research in towns and villages in the Lake Biwa region, the largest lake in Japan, in the

1980s. The group was led primarily by Hiroyuki Torigoe (1944–), Yukiko Kada (1950–), Akira Furukawa (1951–), and Motoji Matsuda (1955–), who eventually published their research outcomes in such edited volumes as Torigoe and Kada (1984) and Torigoe (1989, 1994), among others. In their work, they put forth life-environmentalism's basic stance: that researchers should "put [themselves] into the position of inhabitants who live their everyday lives (*seikatsu*) within the given society" (Torigoe 1989, 19; Chapter 18, 262). Let us first review the historical background of how they came to espouse this basic stance and the epistemological basis behind it.

In the early 1980s when life-environmentalism was first introduced, two environmental discourses were omnipresent in Japan when it came to tackling environmental problems: natural environmentalism, which focused on the preservation of ecosystems by removing human influence, and modern technocentrism, which saw technological advancement as the ultimate solution to environmental problems (Torigoe 1989). Torigoe and his collaborators were unsatisfied with both of these dualistic environmental discourses. They argued that though they may appear to be contrasting perspectives, they are in fact similar because they view humans and nature as separate and incongruous beings.

Along the Lake Biwa shores at that time, people were still using communally managed waterways that were connected to the lake on a daily basis; sustaining their lifeways was therefore directly connected to the conservation of the aquatic environment. Torigoe, one of the founders of life-environmentalism, explains that life-environmentalism, "was born from the struggle to find ways to solve environmental problems" (Torigoe et al. 2018, 524; author's translation) in situations of close daily human-nature interaction.[1] From the point of view of the residents, caring for the "environment" meant taking care of the resources and socio-ecological systems that supported their everyday lives in the community. Natural environmentalism calls for limiting human activities in order to protect nature (i.e., the residents should stop using the lake resources altogether), but such an argument could hardly have gained the support of the residents. Modern technocentrism seeks to control nature by technology (i.e., use concrete and cement to control the lake and waterways), but in so doing, it may destroy local social and environmental practices that have supported residents' everyday lifeways up to that point. In contrast, Torigoe and his collaborators realized that local communities cannot be separated from the surrounding environment in and through which they carry out their everyday lives, and that conserving this relationship was what was most urgently needed.

The establishment of life-environmentalism as a way to offer practical solutions to environmental problems on the ground overlaps with the development of environmental pragmatism, a lively topic in the field of Environmental Ethics in the United States during the 1990s. Environmental Ethics, which gained prominence in response to growing concerns over global environmental problems in the 1970s, sharply criticized proponents of anthropocentrism for attempting to control nature for the benefit of humans (e.g., Naess 1973; Singer

1977). Instead, it championed a variety of alternative views and ideas, such as the intrinsic value of nature, that natural ecosystems must be preserved, and even such radical propositions as, "all life forms must be treated equally." These early propositions in Environmental Ethics were important, but they tended to be too idealistic to be applied in policy making and project implementation; in other words, their practical utility was heavily limited.

Even as Environmental Ethics was becoming an increasingly established academic discipline in the 1990s, frustrations began to mount because it was felt to be closed off and unable to offer practical contributions to society. Environmental pragmatism emerged out of this frustration. Contributors to Light and Katz's volume *Environmental Pragmatism* (1996) argued that Environmental Ethics up until then decidedly sided with non-anthropocentrism over anthropocentrism, and intrinsic value over instrumental value. Rather than confining Environmental Ethics within such a binary axis, Light, Katz, and other contributors to the volume argued that the discipline should adopt more multidimensional approaches that can solve actual environmental problems. This "pragmatic turn" aimed to make a breakthrough in the discursive space of environmental research by overcoming conceptual binary oppositions that proponents of the pragmatic turn argued have no place in the real world. The basis of such a proposition was very similar to what life-environmentalism advocated in the 1980s, so when Japanese social scientists encountered the imported notion of environmental pragmatism in the 1990s, it did not make much of an impact.[2]

Embracing Complex Human-Nature Relationships

From the above, it is clear that in Japan life-environmentalism preceded environmental pragmatism in simultaneously recognizing the limits of nonhuman-centric environmentalism and advancing practical debates that resonated well with local communities. However, many years have passed since the idea of life-environmentalism was first proposed, and there have been significant changes in environmental policies and in public views on environmental problems since then. For example, the 1997 revision of the River Act can be cited as a symbolic inflection point in the history of environmental policy in Japan (River Bureau of the Ministry of Construction 1999). Until then, the River Act was mainly aimed at hydraulic control and water utilization. The revision brought environmental conservation clearly into the scope of the act. By around 2000, rigid natural environmentalism—life-environmentalism's hypothetical enemy—had lost its earlier influence, and ardent modern technocentrism had also come to be seen as anachronistic. Activists who were once enthusiastic natural environmentalists showed an interest in indigenous knowledge, and engineers who had faith in modern technology designed pre-modern water systems. Seeing these changes on the ground, a criticism that life-environmentalism is no longer novel and has little to offer seems valid to some extent (Inoue 2001).

It is also worth pointing out some key changes in the kind of environmental problems that were prominent in Japan in the postwar period. In the postwar history of environmental issues in Japan, one of the major changes is the shift from industrial pollution to urban pollution, or the shift "from pollution problems to environmental problems" (Horikawa 1999, 213), which took place in the 1970s–1980s.[3] However, what is more relevant to our discussion of life-environmentalism is the critical shift that took place in the 1980s–1990s. Up to that point, Japanese environmental conservation focused primarily on protecting wilderness; then, around the 1980s–1990s, conservation efforts began to explicitly target environments regularly or periodically used and managed by people. In Japanese, these managed environments are referred to as "secondary nature" (*nijiteki shizen*).

Satoyama ("village mountain"), which symbolizes such "secondary nature" in Japan, is a rural landscape in which various human-managed elements, such as coppices, paddy fields, reservoirs, vegetable fields, and grasslands, are studded in mosaic patterns (Takeuchi et al. 2003; Takeuchi 2010). Since the 1960s, as people no longer obtained food, fuel, and materials necessary for their daily use from nearby nature, *satoyama* became unmanaged. As a result, there was ecological succession and the number of once-common plant and animal species, such as *medaka* fish (*Oryzias latipes*) and Asian fawn lilies (*Erythronium japonicum*), decreased significantly. Accordingly, some observers began to advocate for the importance of secondary nature, in which humans regularly or periodically intervened, from the standpoint of biodiversity; they noted that secondary nature actually supported not only human livelihoods and production but also rich biodiversity, maybe even more so than primary nature did (Moriyama 1988). This problem of resource *underuse* came to be known as the "*satoyama* crisis." It is now recognized in the National Biodiversity Strategy and Action Plan as one of the four major threats to Japan's biodiversity, along with resource overuse, invasive species/chemical substances, and global warming (Ministry of the Environment 2012).

Resource underuse problems require the restoration of human intervention into nature, and neither natural environmentalism nor modern technocentrism can respond adequately to such problems. The former tends to favor "natural" succession of vegetation, hence not solving the identified problem, and the latter's emphasis on the use of universalizing modernist knowledge tends to discard locally specific human-nature relationships.

In response to the changing nature of environmental problems, there was an increase in research on human-environment relationships through the 1990s in Japan, notably in fields such as environmental folklore studies, environmental anthropology, conservation ecology, and ecological engineering. These studies investigated the conditions of and historical changes in human-nature interactions in specific regions, and sought to take advantage of the wisdom and skills that had been used to manage local environmental resources in the past as a guide for future conservation and sustainability efforts. The rise of such new research programs in the 1990s seems to have further lessened the uniqueness of life-environmentalism in the academy.

I argue, however, that these new developments do not mean that life-environmentalism has come to be fully understood among scholars and policymakers. The distinction between surface-level similarities and core ethos is centrally important here. That is, ideas whose basic principles seem to align with life-environmentalism are indeed more widely professed and policies and projects that draw on similar ideas are increasingly brought into reality, but these initiatives almost never truly grasp the fundamental heart and ethos of life-environmentalism. In order to understand this crucial difference, I now turn to a more thorough analysis of the epistemology and ethos of life-environmentalism.

The Epistemology and Ethos of Life-Environmentalism

A Theory Arising from Fieldworkers' Experience

Since the Earth Summit in 1992, sustainability has become a common agenda globally. One of its most important goals is the protection of local inhabitants' lives and their environment. At first glance, life-environmentalism may seem to fit well with these global ideals. Certainly, if one imagines a utopia where people live in harmony with nature, these global visions and life-environmentalism may appear to be in line. However, it is not by dealing in ideals and visions that a theory shows its true character and strength, but in facing the most challenging of imaginable realities.

Life-environmentalism is an approach born from and rooted in fieldwork (Torigoe 1989). Fieldworkers are often thrown into confusion in the field and experience the destabilization of their worldview. At that time, researchers are tested and can either insist on analyzing the research object by applying an already established cognitive framework, or allow the framework to be overturned by getting even closer to the object. Practitioners of life-environmentalism sought to construct a new approach while affirming the overturning of their worldviews. In this sense, life-environmentalism is indeed a "fieldworkers' theory" (Torigoe 1989, 3), and is fundamentally incompatible with global environmental ideologies that are rooted in the realm of modern enlightenment, which seeks to arrive at universal truths.

This anti-enlightenment tendency of life-environmentalism, which may undermine researchers' own worldviews and positions of authority, is an important feature of the approach. If we follow this line of thinking further, life-environmentalism's claim seems to boil down to that because each local area has its own unique lifeways, there must be a unique way to promote locally specific mixes of development and preservation suited to each place. At this point, upholding such universal ideas as the preservation of nature or the security of inhabitants' lives above all else does not really help on the ground. Accordingly, the next question becomes from what standpoint life-environmentalism views and understands the subject, make judgments about its observations, and approve certain actions and policies over others. To that end, let us now take a closer look at the positionality of those who practice life-environmentalism.

New Philosophy of Science and Impact from Environmental History

When life-environmentalism was introduced, about 20 years had already passed since the influential discussion of Hanson's theory-ladenness (Hanson 1958) and Thomas Kuhn's paradigm theory (Kuhn 1962). A post-Kuhnian philosophy of science had successfully challenged the conventional positivist view of science and its ideas permeated widely into the social sciences in Japan. Hiroyuki Torigoe and his colleagues, influenced by those intellectual currents, as well as by such Japanese scholars of the history of science and science philosophy as Yōichirō Murakami and Shigeru Nakayama, delved into the question of how they could best understand the local communities that they studied. There, they were inspired by the idea of environmental history.

In 1984, Torigoe posed a question about positionality in understanding social phenomena and asserted that there could not be a third person's positionality in environmental issues. For example, he argued that if one looks at the environmental history of westward expansion in the United States from the perspective of the indigenous peoples of America, this was nothing but a history of genocide and destruction (Torigoe and Kada 1984). In this way Torigoe rejected the possibility of a scientifically objective position, but his aim was also to question what constitutes truly meaningful knowledge. That is, for Torigoe, the subject of environmental history is "the accumulated knowledge of the past," which in turn generates "everyday knowledge" different from "scientific knowledge." Moreover, he ascertained that the everyday knowledge of inhabitants is what critically supports their local environment. Thus, studying environmental history by putting yourself "in the position of local residents" is not simply about attaining objective knowledge about traditional interactions of inhabitants and their environment. It is more than this. Researchers must seek to reveal the unique life logics that are underpinned and informed by inhabitants' everyday knowledge (Torigoe and Kada 1984). The value of environmental history when studying local environmental problems is well recognized today, but it was the advocates of life-environmentalism, including Torigoe, who first realized the methodological potential of environmental history and incorporated it into environmental sociological research in Japan.

However, in the 1980s, expressing such a methodological proposition was a challenge to the conventional scientific views of the time. Torigoe looks back on the time and remembers that they had to confront more senior colleagues who held authoritative positions within universities in order to establish their own research stance (personal communication with Torigoe, March 30, 2007). Indeed, Torigoe remembers that they were repeatedly criticized by senior researchers for abandoning the objectivity needed for science. In response, Torigoe, after much thought, developed the argument that life-environmentalism sides with the position of the "life systems" by which inhabitants live, rather than with the inhabitants themselves; accordingly, if individual inhabitants conduct activities that weaken the sustainability of the local life system, they would be subjected to criticism (Torigoe et al. 2018, 526). That is, Torigoe altered the fundamental stance of life-environmentalism, arguing that it does not seek to put itself into "the position of the *inhabitants* who live their everyday lives" (Torigoe 1989, 19; Chapter 18, 262); instead, it seeks to analyze from "the position of *the everyday life (seikatsu)* of the inhabitants" (Torigoe 1997, 26; author's translation and emphasis). This correction may seem to make sense at a glance, and may seem to have successfully escaped severe criticism. But who can know what the everyday activities that sustain everyday life (*seikatsu*) should be like and who can claim to analyze from the standpoint of everyday life (*seikatsu*)?

This question was reflected in the critical comments of sociologist Koichi Hasegawa (1996). Hasegawa charged that if life-environmentalists claimed to speak from the position of everyday lifeways, then they would have to bring some normative value standards (e.g., what ideal relationship between humans and nature *should* be) into the field with them, and then use it to determine whether or not particular practices and cultures of inhabitants meet their standards (Hasegawa 1996). Another sociologist, Kōkichirō Miura, also sharply alleged that life-environmentalism actually relies on deductive theorization; that its claims cannot be contested, criticized, or modified based on empirical data; and that it neglects data that deviate from its theoretical premise and only collects data that match the theory (Miura 2005, 48; also see Miura (1995)). In addition, geographer Shizuyo Sano pointed out that environmental sociological research based on life-environmentalism fails to recognize the importance of temporal changes, lacking long-term historical perspectives (Sano 2006).

Similar criticisms were seen in the field of Japanese folklore studies (*minzoku-gaku*). Some environmental folklore research that is grounded in life-environmentalism identified elements in traditional technologies and folk cultures that contributed to environmental conservation and sought to learn lessons from them for a sustainable society. However, those studies were criticized for conveniently selecting only symbiotic, harmonious human-nature relationships from among the many types of human-nature relationships. Natural disasters, for example, are a kind of human-nature relationship that can hardly be considered harmonious (Shinohara 1994).

Originally, life-environmentalism's key insight, arising from fieldwork, was that it was not possible to take an objective position on environmental problems. Despite this very origin, when Torigoe reformulated its central claim as seeking to "put itself into the position of everyday life (*seikatsu*)," rather than that of inhabitants, he was faced with various well-justified criticisms. Indeed, Torigoe later took back the modification, and returned to the original proposition, that life-environmentalism seeks to "put [itself] into the position of inhabitants" (Torigoe et al. 2018). However, we must note that when Torigoe claims to "put ourselves into the position of inhabitants," it is not exactly the same as the expression of researcher's positionality that is now often seen in Euro-American environmental justice literature and social movement literature (e.g., when researchers say, "we stand by/stand with local residents" in their struggles in the face of injustices). To further clarify Torigoe's and other life-environmentalists' stance, let us dig deeper into what they mean by, "putting [themselves] into the position of inhabitants."

Experience as a Critical Reality

Epistemologically, life-environmentalism does not follow positivism, which assumes that it is possible to objectively grasp social phenomena, similar to how natural phenomena can be objectively observed in the natural sciences. Life-environmentalism argues that rather than observing the actions themselves, which tends to be a traditional focus of sociological analysis, we must descend to the foundation from which the actions were generated. To that end, life-environmentalism chose to focus on "(lived) experience" (*keiken*) rather than "actions" (*kōi*) as the basic unit of analysis (Torigoe 1989). "(Lived) experience" here does not simply refer to past actions. As I understand it, it is the process of people constantly reconstructing those past actions in their inner worlds, as they refer to the present, the future, and beyond (cf. Chapter 12). Lived experience, which one would assume is in the past, is therefore actually continuously recreated in the present, even as the present and imagined future influence how we experience our past.

Torigoe metaphorically presents us with a proposition about how to grasp the experience behind actions: "we cannot understand the mind of a person, but we *can* understand the mind of a group of people" (Torigoe 1989, 45; Chapter 18, 277; emphasis by author). Importantly, "the mind of a group of people" here does not mean some shared consciousness of multiple individuals (in the sense of "what's in everyone's mind"). Rather, the metaphorical statement suggests that it is possible to understand group-based local logics of legitimacy or justification, referred to as *iibun* (Torigoe 2014), because it is usually articulated in order to communicate with and persuade others (at least those within the group). But because "the mind of a group of people"—local logics and locally legitimized ways of understanding and doing things—are formed on the basis of lived everyday knowledge, it needs to be understood from the perspective of the inhabitants rather than from that of objectivism. This is why life-environmentalism does not simply accept manifested actions of inhabitants as "the position of inhabitants." Rather, putting oneself into the position of inhabitants means understanding the group-based local logic that reflects the lived experience of the inhabitants.

Cultural anthropologist Motoji Matsuda offers another explanation about the same proposition (Matsuda 1989). That is, when some common understanding or consensus is formed within a local community, it often becomes necessary for community members to justify the logic that led to consensus formation. Persuading others and themselves is a crucial part of this process. At that time, there are multiple possible discourses for persuasion at their disposal, and community members choose the "most convenient" discourse for the given situation. Fieldworkers often encounter situations where a local person's opinion appears to have suddenly changed, for example, from opposition to approval on a particular state-led project. In such cases, it is useless to focus on expressed opinions (i.e., the act of speaking). Rather, Matsuda argues, it is critical to assume the presence of and grasp the range of latent "discourses of persuasion" behind the apparent action.[4]

Furthermore, Matsuda takes special note of the role played by "(everyday) life knowledge" (*seikatsu-chi*)[5] which steers the "the discourses of persuasion" toward choices that will prioritize and support the practical utility of everyday life. This is because he realized that people exercise such knowledge in the face of external forces of domination—at times choosing to seek compromises with it, at other times choosing to submit temporarily to it, and sometimes to violently resist it (Matsuda 1989, 125). In other words, he finds that those who live in sites of environmental controversies devise methods to deal with external power and to secure their lives by basing their choice of appropriate discourses not on objective reasoning or some exceptionally clear sense of their intentions, but on what works for their everyday life and for the time being. "Putting oneself into the position of inhabitants," therefore, also means to pay careful attention to such "(everyday) life knowledge."

In summary, life-environmentalism has an epistemological characteristic that the inhabitants' standpoint is understood not by interpreting apparent "actions" alone but by descending to their "(lived) experience." This experience needs to be grasped by interpreting a range of different local logics and locally legitimate discourses (*iibun*) that may be revealed in different situations. Which *iibun* is considered appropriate in a given circumstance is dictated by residents' "(everyday) life knowledge." From these explanations, it should be clear that life-environmentalism is, at least in principle, not as simple as extracting observations that match a predefined ideal of human-nature relationships. There is nevertheless a risk of arbitrariness in interpretation, especially arising from researchers' emotional sensitivity. In order to account for such concerns and criticisms, researchers must offer thick descriptions of their own thought process, explaining how they came to understand "the people's mind" and how they validated their interpretation. The validity of their interpretation, based on the premise of life-environmentalism, is ultimately judged by the inhabitants of the study area.

Putting Life-Environmentalism into Practice

Environmental Justice and Inhabitants' Self-Determination

In practice, life-environmentalism is often seen as endorsing residents' self-determination (Horikawa 1999, 217). This interpretation leads us to consider potential connections between life-environmentalism and the study of environmental justice. The idea of environmental justice (EJ) emerged in the 1980s as a critical response to the inequity brought about by the disproportionate environmental burdens imposed on areas with blue-collar residents and racial minorities in the United States. The academic field of environmental justice and the environmental justice movement uphold that socio-economically marginalized people also deserve to enjoy a good environment. EJ also supports efforts by marginalized communities to assert their right to determine their own living environment. As a result, environmental justice movements are often also connected with indigenous peoples' movements.

EJ is critically different from non-anthropocentric environmentalism, such as the theory of rights of nature, the theory of animal liberation, and deep ecology, in that the emphasis is on multiculturalism and on local self-determination. This emphasis has been supported by case studies of small communities with relatively closed material cycles where autonomous decisions made by the inhabitants tend to inherently result in environmental conservation. However, in today's rapidly globalizing world, in which interactions with the outside world are ever more frequent and intense, many community-based organizations and socio-ecological systems that were once autonomous and sustainable have been transformed. For this reason, the autonomous decision-making of local residents does not always result in the conservation of the local environment. Of course, even if we can agree that decisions made by local communities should be prioritized over environmental conservation, it is clear that the concept of environmental justice embodies a latent conflict between local self-determination and environmental conservation (Ishiyama 2004).[6]

This critical observation is also often made about life-environmentalism. Life-environmentalism has declared that it "seeks to put itself into the position of inhabitants" to solve local environmental problems, but why does it seem to almost always view the self-determination of residents so positively? To be sure, it is not because local self-determination always means that the natural environment will be preserved. For example, Torigoe argues that if a given community is located in a low-lying wetland where the residents lament that in the case of a flood "it's a miserable place where even one's casket would float away," then life-environmentalists would approve of modern construction work to enhance drainage (Torigoe [1984]1991, 332; author's translation). They would take this stance because of the "necessities of living" (Torigoe and Kada 1984) and because it is practical/expedient for those who live there (what Matsuda (1989, 117) calls *"seikatsu no bengi"*), both of which are key values of life-environmentalism. Yet, the construction would destroy the habitats of plants and animals that have long lived in the wetland.

Life-environmentalism does not directly respond to such a critique; instead, it sidesteps it and proclaims the conviction discovered through fieldwork: "people still live their lives" (Furukawa 1999, 149; author's translation). Not surprisingly, in cases such as this, life-environmentalism could be accused of justifying material development, which was a mainstay of postwar conservativism in local politics in Japan.

At the time when advocates of life-environmentalism conducted their study on the shores of Lake Biwa, they were able to see how protecting life systems embedded in the local community could lead to the preservation of the natural environment, which was closely integrated into the lives of the people of those communities. Indeed, life-environmentalism was born at a time when a reciprocal relationship between people and nature could still be observed when conducting fieldwork, or when the existence of such a positive relationship in the recent past could at the least still be discerned through interviews with informants. If so, what use is life-environmentalism in contemporary society where such "happy" human-nature relationships have nearly vanished? Pointing out this problem, Hasegawa (1996) asserted that symbiotic relationships between humans and nature may have existed in the past, but even those were exceptions to the rule. He criticized life-environmentalism for being a "romantic retrogression" in the sociology of environmental problems, which, he argued, should become more policy-oriented.

Torigoe had anticipated such criticisms before Hasegawa's critical remarks. To be sure, even though life-environmentalism did arise from sites facing environmental challenges in the Lake Biwa region at a time when the lake environment was being rapidly transformed, practitioners of life-environmentalism did not initiate explicit political or social movements that would affect environmental policy.[7] However, Torigoe and his collaborators thought that, rather than directly participating in political movements, it was essential for them to use environmental history to fully grasp the lifeworld in which people live. That was how they could best contribute to the implementation of what they consider to be "authentic policies (*hontō no seisaku*)" (Torigoe [1984] 1991, iii). They reasoned that if they presented easy-to-understand policy prescriptions, which may be subsequently adopted by the authorities, such policies would only be superficially implemented. Instead, they prioritized the analysis of people's values in order to reveal the essential core of environmental problems. That is why they decided to offer thick descriptions of the rich relationship between people and water in the Lake Biwa shore area, knowing that their work might be read as a cultural study (as opposed to serious work relevant for policy) (Torigoe [1984] 1991).

Torigoe plainly rejected the romantic notion that rural communities are harmonious, idyllic everyday worlds. This is because he had seen many instances in which fissures of "mutual non-understanding" emerged in small communities (see Chapter 18, 265). He clearly saw how these fissures were then exploited by the state and developers, ultimately resulting in deep divides in the community. Still, as Hasegawa pointed out, in such works as Torigoe (1994) and Kada (1995), the proclaimed strategy of intentionally focusing on describing culture seems to have weakened, and those works could be understandably interpreted as "the study of an idyllic and peaceful everyday life" (Hasegawa 1996, 130; author's translation).

It was the students of Torigoe who forcefully countered such criticisms, demonstrating through their empirical studies that life-environmentalism is at its best precisely in sites of environmental struggles where opinions are sharply divided. Case studies include those by Tsuchiya (2018), which focuses on the conflicts over disposal sites of potentially radioactive rubble after the 2011 Great East Japan Earthquake, and Yamamuro (2018), which examines the formation of a new stakeholder community in municipalities around the nuclear facility in Tōkai Village after the 1999 JCO criticality accident (also see Chapter 6). Both of these studies demonstrate that nonlocal actors often demand local residents to clarify their position on a particular issue along a binary line (either for or against), and to participate in social movements according to the binarized controversy. However, these studies show that such nonlocal demands often fail to gain ground in local communities where people carry out their everyday lives.

Such situations can help us to think through the issue of what is meant by "local self-determination" in life-environmentalism. Those who think in terms of "local self-determination" often assume that "self-determination" is about local inhabitants' choice of an answer to a defined problem. In life-environmentalism research, however, one must think through the focal issue from the standpoint of those who have no choice but to cope with a situation imposed on them. From the point of view of life-environmentalists, the scope of local self-determination must include the framing of the problem itself. In order to understand the active agency of inhabitants who may refuse to be interpreted in orthodox frameworks of understanding, researchers will be forced to revise their own worldview and frames of reference. And this is precisely when an intellectual discovery can be made. Only after such struggles in the field might researchers finally be "in the position of inhabitants" and can ideas for realistic possibilities about local cooperation and collaboration with outsiders possibly be envisioned.

Affirming Human Life and Feelings

One may get the impression that life-environmentalism often affirms the choices of inhabitants who try to live their lives even if it means having to destroy their own environment at the scene of an environmental conflict. Faced with the same scene, political ecologists, for example, are likely to critically analyze the macro-level structure of power that created the dire situation. Furukawa (1999) admits that "life-environmentalism lacks a critical perspective on macro-level power," but continues,

> [T]he origin of life-environmentalism is the uneasiness over how critical discourses on macro power [structure] tend to stray from the [mind of] people who live in a given place. Life-environmentalism has consciously emphasized the creativity hidden in the logic of inhabitants, and the rebellious nature of the traditions that they developed without even realizing it. This focus comes as a result of life environmentalists' contention that flawless and rational 'big theories,' however radical and critical theories they might be, have nevertheless always been a variant of modernist epistemology.
>
> (150; author's translation)

In this way, Furukawa justifies his position of putting himself first and foremost into the position of inhabitants, rather than devoting his energy to understanding macro structures with critical theoretical perspectives.

Life-environmentalists' text, including Furukawa's, does not explicitly state that they affirm people's lives. However, I infer that understanding local communities through inhabitants' experiences, by putting themselves in the inhabitants' position, will necessarily mean that they do indeed affirm people's lives. Let me elaborate on this point further.

We humans gain experience through actions and wish to take advantage of that experience for our future actions. By empathizing with others who are also gaining experience in this way and by influencing each other, we wish to live better lives. In life-environmentalism, researchers seek to capture the agency of people who flexibly adjust in order to live well, rather than describe an unswerving way of life unaffected by emotions. Hence, they would not pass judgment on a person's lived life experience just based on the few actions to which it led that they are able to witness. If they were to do so, that would mean denying the very existence of the given people as human beings. This view, which is central to life-environmentalism, seems to be based not so much on trust in the local residents, but more on a sense of affirmation of human life. Therefore, in my view, what distinguishes life-environmentalism from many other approaches is the way that field researchers are first and foremost fascinated by the very existence of those who live in a given community; describe their ways of living, paying special attention to their lived experiences; and then finally think about the ways in which people interact with their environment and their relationships with the systems, both human and natural, that support their everyday living.

Life-environmentalism attends to feelings, or affectivity, which may be a rather difficult concept to properly handle in the social sciences. Surely, people often make judgments and act based on feelings; hence, feelings are an essential element in analyzing society. Psychologists may attempt to employ positivist empirical methods such as electroencephalography (collecting brain wave data) to try to analyze feelings and affects. In my view, life-environmentalism offers another approach to capture the issues of feelings and affectivity.

Specifically, I propose to consider experience and emotional sensitivity as correlated concepts, and to articulate their relations within the experience-based theorization of life-environmentalism. Emotional sensitivity refers to sensitivity and awareness of emotions and feelings. People's life experiences are constantly updated and such updates are accompanied by subtle emotional sensitivity. At the same time, when researchers try to understand the experiences of inhabitants, emotional awareness and response is also required of the researchers.

The heightened emotional sensitivity that researchers try to sharpen in order to grasp the experience of local inhabitants naturally destabilizes the researchers' worldview. This stance, characterized by emotional sensitivity and empathy, does not allow for the study subject to be captured from a third-person objective perspective. It demands that the researcher share in the subjective experience of facing the given issues as a human being, just as the inhabitants. In life-environmentalism, researchers place themselves in the field in order to examine the relationship between people's lives and their environment and thereby learn how the relationship between society and environment, and between humans mitigated by environment, can go smoothly. If researchers do not believe in the power of humans to manage their lives even in the face of great hardship, they will soon be too exhausted to continue their intellectual pursuit. In other words, the affirmation of human life in life-environmentalism seems to be linked to the researcher's own belief that there must be a solution to the specific environmental problems in their given study site.[8]

Toward Authentic Policies

Standing in the sites of environmental problems, Torigoe contemplated the ideal of realizing a kind of "authentic policy," which can only be attained by fully grasping the life worlds and lived experiences of local people from life-environmentalist perspectives (Torigoe [1984] 1991). The idea is that if such ideal policymaking were to ever be implemented, it would mean the attainment of several ideals at once: (1) a society that respects people's lived history and the complex depths of people's everyday lives and (2) the protection of the local environment, which would result from such policymaking and such a society (Torigoe et al. 2018).

However, there is no denying that nowadays, it is difficult to introduce experience and emotional sensitivity, which are the essence of the idea of life-environmentalism, into environmental policy. It is hard for such experience-based and affective arguments to break through because even if attempts are made to include ideas from the social sciences in public policy debates, positivist epistemology continues to dominate over other approaches, as evidenced by approaches such as Evidence-based Policy Making (EBPM) becoming mainstream. Such ideas as experience and affectivity based on interpretive epistemology are looked down upon and neglected in policy making. For this reason, if we want to realize the kind of ideal "genuine policy" envisioned by Torigoe, we must first generate such policies from among the people within the community who desire such a policy. If the policy implemented in the community succeeds in enhancing the well-being of the people, the scope of the policy may be expanded outside the original community (where members share some lived experiences, and where everyday life knowledge is privileged) toward the realm of public sphere (where individual citizens engage in reasoned discussion, and where formal-knowledge is privileged).

Life-environmentalism has focused on "small communities" where face-to-face relationships can be established and maintained (Torigoe 1997; Torigoe et al. 2018). In light of the discussion here, this refers to local communities where shared lived experience and emotional connections (whether they are good, bad, or complicated) play a critical role. Torigoe argues that humans have always lived in families and communities, but while families tend to be "closed" and "private," small communities are "open" to outside, which he considered to be an important quality.

Nevertheless, in recent years, bonds within existing communities have become weaker, and "individuation" has progressed. Is it possible in this modern society to create the kind of "small communities" that life-environmentalism presupposes? If such communities cannot be created intentionally, what conditions are required for such communities to emerge? If life-environmentalism is to engage in policy discussion, I believe that it is imperative first to clarify the conditions required for a small community to come into existence, and then present strategies to expand policies realized within the community to the realm of the public.

Notes

1 "Environmental problems" in postwar Japan until the 1980s typically referred to an array of issues arising from industrial development and modernization, including, but not limited to, water and air pollution and environmental destruction through large-scale construction projects.
2 Moral philosopher Shirouzu (2004) discusses the affinity between the U.S. environmental pragmatism and Japanese life-environmentalism, focusing on the study of *satoyama*.
3 Of course, industrial pollution problems have not disappeared today. Sociologist Saburō Horikawa examined the number of studies dealing with pollution and environmental problems in environmental sociology in Japan, and found the year 1975 to be a turning point from pollution-focused to environment-focused research. However, Horikawa carefully sums up this change with the phrase "environmental research in addition to pollution research" (Horikawa 1999, 213).
4 Editors' note: Let us give a more concrete example to illustrate the significance of these latent discourses of persuasion, which Matsuda calls "idioms." Think of a situation in which a large waste treatment facility in a local community. Residents of the community are likely to respond to the proposal in various, often conflicting ways. Some would be for the project, saying "we should cooperate with this for public welfare," while others may oppose, arguing, "this will be a source of pollution" or "the values of surrounding properties will get a hit." Moreover, some of the opposing residents may alter their position, saying "our relatives begged us (to change our mind)." Yet, others may try to persuade their acquaintances by saying, "the village is one." In this case, such phrases as "for public welfare," "pollution," "impact on property values," "(words of) relatives," and "the village is one" are all seen as (mere) idioms, which are rooted in the experiences of the community, and are frequently mobilized in persuading others and themselves. Matsuda emphatically argues, however, it is critical not to give a particular idiom an excessive meaning, or to associate it too deeply with the speaker's attributes. For example, one does not need to, and should not, infer "he expressed concerns over property values because he is a big landholder in the region," which may lead the researcher to analyze unequal socio-economic structure of the community. In short, life-environmentalists consider the range of idioms that circulate in a local community to be reflective of the community members' lived experience, but resist to interpret each individual idiom as the expression of some essential quality of the community or the individual.
5 Editors' note: Elsewhere Matsuda uses the term *bengi-chi* ("expedient knowledge") to describe this idea.
6 Geographer Noriko Ishiyama examines a case where the Goshute tribe in Skull Valley in Utah decided to host a nuclear waste facility, and pointed out the flaws in the distributive justice theory that prevailed in early environmental justice research. She argues for the need to deepen the theory of procedural justice that incorporates the concepts of self-determination and political ecology (Ishiyama 2004).
7 Hasegawa's criticism was originally written as a book review of Kada (1995). Kada later became the governor of Shiga Prefecture, where Lake Biwa is located, and served for two terms (2006–2014).
8 Another recent approach that takes a similar position in Japanese environmental sociology is the literature on adaptive governance (Miyauchi 2013, 2017). In his discussion of environmental governance, Taisuke Miyauchi states, "the adaptive governance for which we advocate is the way of governance that trusts in society's ability to adapt" (Miyauchi 2013, 27–28). Miyauchi, like life-environmentalists, thinks that local environmental problems will never be solved if the local community itself is excluded from discussion.

References

Furukawa, Akira. 1999. "Kankyō no Shakaishi Kenkyū no Shiten to Hōhō–Seikatsu Kankyō Shugi toiu Hōhō" [Perspectives and Methods of Environmental Social History Research: A Method of Life-environmentalism], In *Kankyō Shakaigaku Nyūmon–Kankyōmondai Kenkyū no Riron to Gihō* [Introduction to Environmental Sociology: Theories and Techniques of Environmental Problems Research], edited by Harutoshi Funabashi and Akira Furukawa. 125–52. Tokyo: Bunka Shobō Hakubunsha.

Hanson, Norwood Russell. 1958. *Patterns of Discovery: An Inquiry into the Conceptual Foundations of Science*. Cambridge: Cambridge University Press.

Hasegawa, Koichi. 1996. "Shohyō Kada Yukiko cho–*Seikatsu Sekai no Kankyōgaku*" [Book Review: Kada, Yukiko "Environmental Studies in the Living World"]. *Soshioroji* [Sociology], 41 (2): 128–31.

Horikawa, Saburō. 1999. "The Rise and Institutionalization of Environmental Sociology: An Overview and Assessment of the Japanese Experience, 1945–1998." *Journal of Environmental Sociology*, 5: 211–23. In Japanese.

Inoue, Takao. 2001. *Gendai Kankyō Mondai Ron: Riron to Hōhō no Saiteichi no Tameni* [Reformulation of Theory and Method on Environmental Problems]. Tokyo: Toshinsha.

Ishiyama, Noriko. 2004. *Beikoku Senjūminzoku to Kakuhaikibutsu–Kankyō Seigi o Meguru Tōsō* [American Indigenous Peoples and Nuclear Waste: The Struggle for Environmental Justice]. Tokyo: Akashi Shoten.

Kada, Yukiko. 1995. *Seikatsu Sekai no Kankyōgaku–Biwako kara no Messēji* [Environmental Studies of the Living World: A Message from Lake Biwa]. Tokyo: Nōsan Gyoson Bunka Kyōkai (Rural Culture Association Japan).

Kuhn, Thomas. 1962. *The Structure of Scientific Revolutions*. Chicago: University of Chicago Press.

Light, Andrew, and Eric Katz eds. 1996. *Environmental Pragmatism*. London: Routledge.

Matsuda, Motoji. 1989. "Histuzen kara Bengi e – Seikatsu Kankyō shugi no Ninshikiron" [From Inevitability to Expediency: Epistemology of Life Environmentalism]. In *Kankyō Mondai no Shakai Riron: Seikatsu Kankyō Shugi no Tachiba Kara* [Social Theory of Environmental Problems: From the Standpoint of Life-environmentalism], edited by Hiroyuki Torigoe, 93–132. Tokyo: Ochanomizu Shobō.

Ministry of the Environment. 2012. "Seibutsu Tayōsei Kokka Senryaku 2012–2020–Yutakana Shizen Kyōsei Shakai no Jitsugen ni Muketa Rōdomappu" [National Biodiversity Strategy of Japan 2012–2020: Roadmap Towards the Establishment of an Enriched Society in Harmony with Nature]. Accessed February 22, 2021. https://www.biodic.go.jp/biodiversity/about/initiatives/files/2012-2020/01_honbun.pdf

Miura, Kōkichirō. 1995. "Kankyō no Teigi to Kihanka no Chikara–Naraken no Shokuniku Ryūtsū Sentā Kensetsu Mondai to Kankyō Hyōshō no Seisei" [Normative Power and the Definition of the Environment: Meat Distribution Center Construction Problems in Nara Prefecture and the Creation of Environmental Representation]. *Shakaigaku Hyōron* [Sociological Review], 45: 469–85.

———. 2005. "Environmental Hegemony and Structural Discrimination." *Journal of Environmental Sociology*, 11: 39–51. In Japanese.

Miyauchi, Taisuke. 2013. *Naze Kankyō Hozen wa Umaku Ikanai noka–Genba kara Kangaeru "Junnōteki Gabanansu" no Kanōsei* [Why Environmental Conservation Doesn't Work: The Possibility of "Adaptive Governance" from the Field]. Tokyo: Shinsensha.

———. 2017. *Dōsureba Kankyō Hozen wa Umaku Iku no ka–Genba kara Kangaeru "Junnōteki Gabanansu" no Susumekata* [How Can Environmental Conservation

Work?: How to Promote "Adaptive Governance" from the Field's Perspective]. Tokyo: Shinsensha.
Moriyama, Hiroshi. 1988. *Shizen o Mamoru towa Dōyū Koto ka* [What Does "Preserving the Nature" mean?]. Tokyo: Nōsan Gyoson Bunka Kyōkai (Rural Culture Association Japan).
Naess, Arne. 1973. "The Shallow and the Deep, Long-Range Ecology Movement. A Summary." *Inquiry: A Journal of Medical Care Organization, Provision and Financing* 16 (1–4): 95–100. https://doi.org/10.1080/00201747308601682
River Bureau of the Ministry of Construction. 1999. "The River Law with Commentary by Article: Legal Framework for River and Water Management in Japan." Infrastructure Development Institute. http://www.idi.or.jp/publication/
Sano, Shizuyo. 2006. "A Research Perspective of Environmental History in Japan, with Special Reference to Research on Traditional Subsistence Activities and Landscapes." *The Shirin/Journal of History*, 89 (5): 743–70.
Shinohara, Toru. 1994. "Kankyō Minzokugaku no Kanōsei" [Potential of Environmental Folklore]. *Bulletin of the Folklore Society of Japan*, 200: 111–25.
Shirouzu, Shiro. 2004. "Kankyō Puragumatizumu to Aratana Kankyō Rinrigaku no Shimei–'Shizen no Kenri' to 'Satoyama' no Saikaishaku e Mukete" [Environmental Pragmatism and the Mission of Environmental Ethics: Reinterpreting Natural Rights and the Development of Woodland Areas]. In *Iwanami Oyō Rinrigaku Kōgi 2 Kankyō* [Iwanami Applied Ethics Lecture no.2: Environment], edited by Mitsugu Ochi et al, 160–79. Tokyo: Iwanami Shoten.
Singer, Peter. 1977. *Animal Liberation*. Book, Whole. New York: Avon Books.
Takeuchi, Kazuhiko. 2010. "Rebuilding the Relationship between People and Nature: The Satoyama Initiative." *Ecological Research* 25 (5): 891–97.
Takeuchi, Kazuhiko, Robert D. Brown, Izumi Washitani, Atsushi Tsunekawa, and Makoto Yokohari, eds. 2003. *Satoyama: The Traditional Rural Landscape of Japan*. Tokyo: Springer Japan. https://doi.org/10.1007/978-4-431-67861-8
Torigoe, Hiroyuki. [1984] 1991. "Hōhō to Shite no Kankyōshi" [Methods for Environmental History]. In *Mizu to Hito no Kankyōshi–Biwako Hōkokusho: Zōho Ban* [Environmental History of Water and People–Lake Biwa Report: Expanded Edition], edited by Hiroyuki Torigoe and Yukiko Kada, 328–47. Tokyo: Ochanomizu Shobō.
———. 1997. *Kankyō Shakaigaku No Riron to Jissen: Seikatsu Kankyō Shugi No Tachiba Kara* [Theory and Practice of Environmental Sociology: Perspectives of Life Environmentalism]. Tokyo: Yūhikaku.
Torigoe, Hiroyuki, ed. 1989. *Kankyō Mondai no Shakai Riron: Seikatsu Kankyō Shugi no Tachiba Kara* [Social Theory of Environmental Problems: From the Standpoint of Life-environmentalism]. Tokyo: Ochanomizu Shobō.
———. 1994. *Kokoromi Toshite no Kankyō Minzokugaku–Biwako no Firudo kara* [Environmental Folklore: From the Field of Lake Biwa]. Tokyo: Yūzankaku.
Torigoe, Hiroyuki, Adachi, Shigekazu, and Kanebishi, Kiyoshu, eds. 2018. *Seikatsu Kankyō Shugi no Komyuniti Bunseki* [Community Analysis of Life-Environmentalism]. Kyoto: Minerva Publishing.
Torigoe, Hiroyuki, and Yukiko Kada, eds. 1984. *Mizu to Hito no Kankyōshi: Biwako Hōkoku Sho* [Environmental History of Water and People: Lake Biwa Report]. Tokyo: Ochanomizu Shobō.

Part VI

Translated Excerpts from the *Sociological Theory of Environmental Problems* (1989)

18 Original Introduction of Life-Environmentalism (1989)

Hiroyuki Torigoe

This chapter provides the translation of one of the first texts that offers a systematic introduction to life-environmentalism. The text was originally published as part of the book titled *Kankyō Mondai no Shakai Riron: Seikatsu kankyō Shugi no Tachiba kara* [Sociological Theory of Environmental Problems: From the Standpoint of Life-Environmentalism), edited by Hiroyuki Torigoe in 1989. The book features contributions from members of the Lake Biwa research team, representing various disciplines. Their perspectives and understanding of what has come to be known as life-environmentalism are not necessarily uniform; some of the claims and assumptions may be no longer novel or valid, and there have been modifications to ideas and propositions over the years as well. Nevertheless, the book remains an important reference for both proponents and critics of life-environmentalism, making it worth including in the original form. Specifically, this chapter includes the preface of the book (pages 3–11) and the opening chapter (pages 13–53), both of which were written by Torigoe, the principal investigator of the research team.

The Position of Life-Environmentalism—Preface (pp. 3–11)

Theory Constructed by Fieldworkers

The term "theory" is now used in a wide variety of ways in the social sciences. Some social scientists prefer to use it in a restricted sense, referring only to a set of propositions (especially causal laws). Some researchers use it slightly more broadly, including models and paradigms as theories. Some even consider an investigation of conceptual categories, like the meaning of "urbanization," as part of theory. Furthermore, some researchers consider any abstracted logic to be a theory. There is no single correct understanding of theory among these perspectives. The definition of "theory" itself will vary depending on the purpose of the researcher.

For fieldworkers like us, "theory" is metaphorically comparable to the grammar[1] we learn when studying a foreign language. It should be useful in helping us understand the subject more precisely, persuasively, and logically. If theory becomes less useful from this perspective, it should be modified. I do not consider theory to be scientifically true, nor do I assign absolute value to mastering it precisely. No matter how well we grasp the grammar of a foreign language, we cannot claim to have completely mastered the language itself.

However, just as good grammar is useful for foreigners, good theory is essential for fieldworkers like us. Just as good grammar can accurately express certain basic characteristics of the people who speak the language, sometime theories of fieldworkers can beautifully portray the characteristics of the subject.

One might argue that a theory of fieldworkers is more accurately described as an analytical framework. However, as becomes evident below, this book goes beyond mere discussion of an analytical framework and extends into the realm of what could be called social philosophy. That is because we believe that realm is still part of the "grammar" for our purposes.

Our theory is not an attempt to revive the theories of classical social thinkers such as Marx and Durkheim, nor is it a systematic introduction to the latest social thinkers. I believe that those efforts aim to overcome the impasse of modernity, which is their strength. We are also similarly searching for a novel recombination of knowledge. However, instead of learning from great thinkers, we strive to transcend current social and social scientific limitations through the study of real, ordinary people's lives. In other words, we aim to create a logic for the recombination of knowledge within the field. This book serves as a milestone of such efforts.[2]

The Effectiveness and Limitations of Life-Environmentalism

Broadly speaking we recognize three different perspectives on environmental change, which we delve more deeply into later in the book. The first is "natural environmentalism," the second is "modern technocentrism," and the third is what we call "life-environmentalism." Each perspective (*shugi*)[3] focuses on nature untouched by humans, modern technology, and people's lives, respectively, and informs policy arguments.

While we clearly stand on the position of "life-environmentalism" in this book, we also recognize that it is not a completely unprecedented perspective. Social scientists who analyze human behavior and humanities scholars who emphasize human thoughts often share positions similar to "life-environmentalism" to some extent. Conversely, those with a background in the natural sciences tend to lean toward "natural environmentalism" or "modern technocentrism."

In part because researchers and project managers involved in environmental issues predominantly come from the natural sciences in contemporary Japan, "natural environmentalism" and "modern technologicalism" hold significant influence as two competing perspectives. Moreover, due to the prevailing belief that science is always objective, these perspectives are rarely scrutinized as ideologies, leading to instances where policy measures are implemented while contradictions between the two perspectives remain hidden.

Our argument for "life-environmentalism" as a perspective (*shugi*) is based on the specific case of Lake Biwa. Typically, the upper reaches of rivers are sparsely populated, but in the case of the Yodo River, which runs through the Keihanshin region [spanning Kyoto, Osaka, and Kobe], the upper reaches include Lake Biwa and several cities, including Ōtsu City. If we hypothetically remove residents in the middle and upper reaches of Yodo River, the river water will become clean. However, in the current situation, where *people have little choice but to live there*, the question is what realistic policies can be established. This question forms the basis of "life-environmentalism."

Relying solely on "modern technocentrism," which prescribe concrete lining of streams and road widening, would be problematic (although civil engineers today claim to be more cautious about environmental impacts and attentive to human lives, the construction of underdrains for road expansion continues). On the other hand, "natural environmentalism," based on the logic of natural ecology, can be overly idealistic in such regions. It is for these reasons that we have developed the concept of "life-environmentalism" as an ideal type (*idealtypus*) (not in the sense of an ideal). However, "life-environmentalism" is not simply positioned between the other two perspectives, and this will become evident throughout the book.

Moreover, it is not that "life-environmentalism" is always superior to the other perspectives. I believe that "life-environmentalism" is useful not only in the Keihanshin region but also in other densely populated areas of Japan and other Asian countries, but there are many instances where "natural environmentalism" is more effective in areas with very low population density. Of course, there are also conditions under which "modern technocentrism" is useful. However, our strong advocacy for "life-environmentalism" stems from the fact that this way of thinking has been downplayed due to the lack of sufficient logical formalization, despite elements of this thinking being expressed by those who interact with the local environment.

Having said that, in this book, we do not provide a general or enlightening introduction to demonstrate the utility of "life-environmentalism." Instead, we focus more on the social scientific logical structure based on this *shugi*. We refer the reader to *Mizu to Hito no Kankyōshi: Biwako Hokokusho* (The Environmental History of Water and People: A Report on Lake Biwa) edited by Torigoe and Kada (1984), a sister volume to this book, for more general and enlightening empirical studies. That book addresses many case studies on the occurrence of environmental problems in Lake Biwa in order to describe and understand the environment as seen from the standpoint of people's lives.

Worm's Eye and Bird's Eye

If we categorize theory into two types, the "worm's eye" and the "bird's eye," the theory generated from concrete field research experiences in this book belongs to the worm's eye. We examine general issues through specific cases that are directly in front of us. This stands in contrast to the bird's-eye view, which considers the dynamics of the entire Earth.

The bird's-eye view is typically found in ideas put forth by international organizations like the United Nations. These organizations propose and implement specific actions, such as "over 100 atmospheric observation stations worldwide should be established." The United Nations refers to these proposals as action plans, which require underlying philosophies or foundational principles. In the 1980s, international organizations addressing environmental issues, including the United Nations, began formulating new foundational principles and policies. The key term at the center of these discussions was "sustainable development."

Currently, the term "sustainable development" is almost universally translated as *jizokuteki kaihatsu*, but those encountering the term for the first time may not fully grasp its meaning. The term could be also translated as *hozenteki kaihatsu* [translator's note: "conserving development"]. In essence, it signifies pursuing development without destroying the natural environment and living environment of the area. Yet, if that were all, the foundation of this key term would be too superficial. This is because achieving both development and natural conservation simultaneously has been considered nearly impossible in reality, no matter how appealing it may sound.

Hence, it is crucial to reflect on how this key term came about. The United Nations Conference on the Human Environment, held in Stockholm in 1972, is renowned for bringing environmental protection to the forefront. Of course, there was a very long history of the environmental protection movement prior to that, but it still holds great importance that global-scale environmental issues were discussed within the United Nations, an organization encompassing countries with conflicting interests in development, including industrialized and developing countries, the global North and South, and large and small countries.

Two primary issues were addressed at the Stockholm Conference. The first concerned the unrestricted use of the Earth's natural resources, while the second focused on the potential risks arising from the rapid growth in population and economic development. These issues were rooted in the fundamental understanding prevalent in the traditional environmental protection movement that conservation and development were inherently conflicting (Documents for the U.N. Conference on the Human Environment 1972).[4]

However, a new way of thinking emerged in the 1980s, suggesting that development and conservation could be complementary. For instance, in situations where extreme poverty leads to the excessive use of agricultural land and subsequent desertification, the environment necessary for supporting development is being destroyed. The underlying logic is that environmental degradation hampers economic growth. Therefore, the argument posits that countries in the early stages of industrialization can achieve full economic growth only when their environmental foundations are adequately preserved.

This line of thinking culminated in the establishment of the United Nations Special Commission on the Environment in 1982, which was originally proposed by Japan. In 1987, the Commission published a report titled *Our Common Future*, which introduced the concept of "sustainable development" as its central idea. Criticisms have been raised regarding the notion of sustainable development, for example, that "it is a compromise between the 'environmental protection' advocated by affluent nations and the 'priority given to development' demanded by poorer nations" (Asahi Shimbun, May 11, 1987, editors' translation). Nevertheless, the impact of the report was undeniable, as evidenced by countries like the United Kingdom, Canada, Nigeria, and the Soviet Union addressing environmental issues in their general speeches at the 43rd session of the United Nations General Assembly in October 1988—an outcome the press referred to as "letting a hundred flowers bloom on environmental issues." [5]

We cannot ignore the movements guided by a bird's-eye view. At least in the 1980s, the concept of "sustainable development" held a certain level of realism and persuasive power. The report *Our Common Future* was particularly influential, and this line of thinking may eventually intersect with "life-environmentalism" in some manner. However, in this book, we intentionally distance ourselves from the prevailing bird's-eye view and instead concentrate on developing internal logical frameworks from a worm's eye perspective. This decision stems from our belief that the theory derived from this standpoint will effectively expose the inherent limitations of the bird's-eye view in the near future. In the future, we will probably raise fundamental questions regarding the meaning of "development" as perceived from a bird's eye perspective. However, that is our future task, and lies beyond the scope of this book. In other words, this book confines itself to a theoretical perspective solely based on a worm's-eye view.

Experience and Life-Environmentalism (Chapter 1, pp. 13–53)

Problems Under Concern and Basic Positions

Environment and Social Organizations

At some point in history, humans began to yearn for change and a "better life." It seems that we are no longer satisfied with a life where we inhabit a stable ecosystem, living as an integral part of the natural world, as perhaps practiced by some hunter–gatherer societies.

This marked the beginning of civilization and the commencement of systematic alterations of the natural environment by humans. These alterations involve multiple individuals utilizing social structures to modify the environment. The scale and nature of these alterations vary, depending on the people involved and the social structures in place. People also organize movements to effectively prevent large-scale environmental deterioration and alteration. Therefore, in order to protect the environment, it is crucial to not only rely on knowledge from the natural sciences but also delve into the complexities of social organizations.

Currently, in Japan, developers typically *make efforts* to obtain the understanding and consent of residents when undertaking alterations to the local environment, as outlined by the City Planning Act and other regulations. However, since the laws do not strictly *require* consent, it is not uncommon for development to proceed with only superficial explanations provided to residents. Nevertheless, persistent and strong public opposition can cause significant consequences. First, it can lead to substantial economic losses for developers and damage to their credibility. Second, the government agencies responsible for issuing permits (such as prefectures and municipalities) may face scrutiny from residents and pressure from certain assembly members who support the residents' concerns. Consequently, it should be rather unusual for environmental alterations to proceed in the face of widespread opposition from residents.

In fact, what is more common when environmental alternation is imminent is a division of opinions among the residents themselves (here "indifference" is regarded as an opinion). Residents are the ultimate and the most important stakeholders in environmental alteration. However, if the residents are divided and unable to reach a unified opinion, their ability to act as a barricade against environmental alterations becomes essentially nonexistent. This is why we must conduct a comprehensive analysis of divisions among residents.

The definition of "residents" varies depending on the relevant laws. Some laws provide a broad definition of residents as "interested parties," allowing for a range of interpretations, while others have stricter definitions such as "inhabitants of the project site and its surrounding areas." However, in contemporary usage, when the term "residents" is mentioned in contexts like "explanations to residents" or "the consent of the residents," it commonly refers to the local *jichikai* [neighborhood association]. Therefore, I adopt *jichikai* as the basic unit of the analytical framework.

Given the aforementioned characteristics, let us examine environmental problems with a focus on the issues of social organization. Compared to the progress made in natural scientific research on environmental problems, there is a significant dearth of research in this area. Thus, we must begin the discussion from a foundational standpoint.

Social Science and Emotional Sensitivity

In the social sciences, such as sociology, that directly deal with human behavior, one of the most formidable problems is the emotional sensitivity of humans. Emotional sensitivity is difficult to fathom, but it is evident that people's judgments and actions are often influenced by it. Actions driven by emotional sensitivity are something that anyone experiences when dealing with other people, such as romantic partners, spouses, or bosses, and they are far from logical. Consequently, the simplest approach is to assume that humans act rationally and that emotional sensitivity does not exist.

Most economic theories, from classical economics to modern economics,[6] are built on this assumption. The human being assumed by economics is called *homo economicus*—someone who acts based on economic rationality. In reality, no human being acts solely based on economic rationality, but by making this assumption, logically consistent economic theories can be formulated.

Another approach is to consider emotional sensitivity as a black box. This approach acknowledges the existence of emotional sensitivity and incorporates it into the logical framework, but does not actually analyze its specific content. Weber's well-known categorization of four types of social action can be seen as belonging to this black box approach.

The four types of social action are based on the motives behind human actions and include: (1) rational purposeful action, (2) value rational action, (3) affective action, and (4) traditional action. The third type, affective action, corresponds to the emotional reactions associated with emotional sensitivity. The first type, rational purposeful action, involves selecting the most suitable means to achieve one's own goals. The second type, value rational action, is guided by beliefs or a sense of duty. The fourth type, traditional action, consists of habitual and repeated actions (i.e., lacking a conscious motive of its own). Therefore, it is somewhat unclear whether the last type can be definitively categorized as an action driven by specific motives.

The first and second types of social action are indeed characterized as rational acts. As a result, there have been numerous sociological empirical studies and theoretical investigations focusing on these two actions. This is primarily due to the fact that rational purposeful action is exemplified by the economic activities found in modern capitalist societies, which often involve elements of value rational actions related to notions of duty, dignity, and trust (although Weber points out that value rational actions are not as "rational" as rational purposeful action). Consequently, these two actions are frequently and extensively subjected to "empirical analysis" (Weber [1922]1972:12–13, translated 39–42). On the other hand, affective action has been acknowledged as a category of social action, but it has received little theoretical development, presumably because it has been treated as a black box.

From Action-Based Approach to Experience-Based Approach

Sociology generally considers action as the basic unit of sociological analysis. This is because social activities originate from actions. There are various theories concerning actions, and Weber's classification of actions, as mentioned earlier, is especially well known. This approach, which categorizes human actions based on motives, exemplifies a science that seeks objectivity by delving into the realm of human subjectivity known as motives.

I believe that many social issues can indeed be addressed effectively from this action-based perspective, which represents a traditional sociological standpoint. However, when it comes to the types of environmental problems we are concerned with, I do not find the action-based approach (*kōi-ron*) particularly effective. Therefore, I opt for the standpoint of the experience-based approach (*keiken-ron*). Before delving into an explanation of this seemingly familiar, yet somewhat unusual term, it is appropriate to present our fundamental ideas about environmental problems.

Three Perspectives (Shugi)

The ideas we present below exhibit certain characteristics that may not align with the dominant modes of social scientific thinking, but we do not intend to be eccentric. There are two primary reasons for this apparent departure. First, we are fieldworkers rather than theorists confined to the study. Second, it is because the subject at hand is environmental problems, which in some ways emerge from the contradictions within modern science.

Let's begin with the issue of the position of scientists, who traditionally strive for objectivity. From the perspective of the history of science, this objective stance distinguished scientists from policy makers and moralists. In principle, this objective stance should be highly valued. However, in reality, every researcher holds some form of ideology, and thus we believe that there is no purely objective standpoint in the field [where environmental problems arise]. Consequently, we have chosen to put ourselves into the position of the inhabitants who live their everyday lives (*seikatsu*) within the given society. By emphasizing the "life" (*seikatsu*) of these inhabitants, we have adopted the standpoint of "life-environmentalism." Identifying ourselves as proponents of a specific "ism" (*shugi*) implies the existence of other -isms as well. When contemplating environmental problems, we can propose two other perspectives (*shugi*) as ideal types.

The first standpoint is the natural ecological standpoint, which considers untouched nature to be the most desirable. Let's refer to this as "natural environmentalism."[7] In this ideology, the focus is not on human life but on nature, with humans regarded as part of the natural world. Although this natural environmentalism is a persuasive perspective, it is criticized from the standpoint of "life-environmentalism" as embodying a "natural science museum vision."[8]

The other standpoint is that of "modern technocentrism." In the context of the hydrological environment, this ideology encompasses large-scale projects such as the "river-basin sewage system," which consolidates all drainage water from a given basin in one location, and the idea of straightening rivers and concretizing their banks and beds. While these ideas are often presented as measures to protect residents, they frequently result in the destruction of their lifeways.

It is difficult to find a researcher whose mind is completely devoid of preconceptions. Most likely, their views align with one of these three broad categories (or a compromise between them). By "ideology" we mean such loosely defined viewpoints. However, it is not uncommon to come across researchers and research-oriented administrators who believe in their own "objectivity" simply because they are engaged in "research," even though they hold a specific ideology. As a result, they tend to diagnose environmental problems without adequately considering their own ideological biases. That is why we feel it is crucial to deliberately emphasize this point.

Experience in Daily Life

Life-environmentalism places conceptual emphasis on "life," or, more specifically, on how we understand "life." There are three distinct characteristics in our approach. The first is the focus on "experience," which serves as the methodological foundation. The second characteristic involves the consideration of "power," and the final characteristic relates to the lack of mutual understanding among human beings.

I will first discuss "experience." As mentioned earlier, life-environmentalism adopts an experience-based approach, rather than an action-based approach. Allow me to explain this further. Typically, what we can observe are the tangible actions of individuals. To understand the reasons (i.e., motives) behind their actions, one can ask them questions using survey instruments or infer their reasons from various other circumstances.

For example, if there is a movement against the construction of a ski resort on a mountain slope, the reasons (motives) for the opposition can be found by a quite simple procedure. Reasons can be classified, or the correlations, such as those between reasons and occupations, can be identified. This type of investigation can be conducted in great detail and for numerous factors, depending on the available resources of funding and time. Even the attributes of the actors alone can be almost indefinitely expanded to include such factors as occupation, age, gender, educational background, place of residence, years of residency, and family structure. In addition, for comparative purposes, one may also include those who did not participate in the movement, allowing for a truly "detailed" study.

This type of study is useful because it allows us to avoid deeply considering our own standpoint or ideology, and because it is generally considered objective. Indeed, we do use this type of survey as well.

However, when I conducted this type of interview survey in the past, I started to wonder about the following. One interviewee expressed his opposition to a particular matter and acted accordingly, but his words and actions did not seem to be particularly steadfast; rather, they may have been accidental outcomes (someone's reasoning resonated with his own thinking, his friend happened to belong to the opposition group, etc.). In other words, what might actually have happened was that he made a choice among several other options at his disposal. The option he chose becomes his expressed opinion, but the other possible options remain unknown. Moreover, these other options may surface when circumstances change (we have encountered too many times the situation in which residents suddenly change their opinion in the midst of an issue). This means that if we only observe an action that emerged as a result of an "accidental" choice, we may develop an irreparable misunderstanding about the research object. That is what I came to realize.

Let us now turn to the issue of the choice of actions. An action is directed toward a certain object/counterpart. Actions related to environmental issues can be characterized as a movement in the broadest sense, and they need to be understood not only in relation to one's own motives but also in consideration of the character of the counterpart. This is where the issue of emotional sensitivity, which I mentioned earlier, becomes a crucial element.

In other words, when a person takes action toward another person, it is normal for the person initiating the action to consider the emotional sensitivity of the recipient. If the recipient does not have emotional sensitivity toward certain actions, the person initiating the action will have to discard options that are beyond the recipient's emotional sensitivity. This is because those actions would lack effectiveness. As the saying goes, it is like "chanting a prayer to the Buddha in a horses ears" [meaning it would be futile]. On the other hand, if someone has strong sensitivity toward a particular matter, the person initiating the action often takes advantage of it to persuade that person. In this case as well, the choice of action becomes "distorted."

When dealing with complex problems within the realm of environmental controversies, at the least this level of analysis is essential; otherwise, one may end up with a ridiculously simplified outline of the problem. Here I would like to refer to what forms the basis of the choice of actions as "experience." That is, I perceive a risk in using "action," which is merely one option among several alternatives, as the basic unit of analysis. This is why this book adopts an experience-based approach rather than an action-based approach.

The assertion that one must descend to the realm of experience in order to analyze the consequences and future possibilities of human actions is not particularly groundbreaking. Whether consciously or unconsciously, fieldworkers have frequently made this assertion. Thus, with a purposeful search, one can easily find similar claims. However, there are few instances where this assertion has been elevated to the level of methodological discussion. To illustrate this point, let us consider the words of a renowned scholar, G.W. Allport, a social psychologist, who states:

> Suppose we take John, a lad of 12 years, and suppose his family background is poor, his father was a criminal, his mother rejected him, his neighborhood is marginal. Suppose that 70 percent of the boys having a similar background become criminals. Does this mean that John himself has a 70 percent chance of delinquency? Not at all. John is a unique being, with a genetic inheritance all his own; his life-experience is his own. His unique world contains influences unknown to the statistician, perhaps an affectionate relation with a certain teacher, or a wise word once spoken by a neighbor. Such factors may be decisive and may offset all average probabilities. There is no 70 per cent chance about John. He either will or will not become delinquent. Only a complete understanding of his personality, of his present and future circumstances, will give us a basis for sure prediction.
>
> (Allport 1962: 411–412)

The above explanation, given Allport's background as a psychologist, naturally appears to sociologists as leaning toward a psychological perspective. However, it aligns with my assertion that predictions of John's future behavior should focus on his experience as a whole. At this point it is necessary to offer a more precise definition of "experience." The term "experience" refers to the one and only experience of each unique individual. Allport uses the term "life-experience," and experience here can be described as such (it is unrelated to *keiken shugi*, which is a conventional translation of "empiricism"). Experience refers to the accumulation of the remembered time from the past for a person or sometimes an institutional body. This is why we used the term "history" in the title of our previous book (editors' note: *The Environmental History of Water and People: A Report on Lake Biwa*). By making experience, rather than action, the basic unit of analysis, time emerges as an important element.

The experience-based approach considers emotional sensitivity as a crucial factor in the choice of action. The emphasis on emotional sensitivity did not stem from a theoretical requirement for logical consistency (if that were the case, we would treat all four [types of actions and motivations] equally as Weber's typology does). Rather, it arises from the reality encountered in the field. Just as Allport draws on examples such as "an affectionate relation with a certain teacher" and "a wise word once spoken by a neighbor," which are closely tied to emotional sensitivity, humans are frequently moved by these factors.

In addition, deep divides, which I call mutual non-understanding,[9] are a common occurrence in everyday life. Reality teaches us that emotional sensitivity is a primary cause of mutual non-understanding. I do not disregard the fact that differences in purpose (rational purposeful) and values (value rational) can also lead to serious conflicts. However, above all else, emotional sensitivity often causes the lack of mutual understanding in reality. In addition, it is surprising how often people's purposes and values originate from their emotional sensitivity. For example, it is not uncommon for individuals to start with the feeling that "nuclear power plants are frightening" and then construct a rational logic to align with the sentiment. For this reason, too, I believe that a strong focus on emotional sensitivity is essential.

Moreover, implied by the term "mutual non-understanding," an analysis based on the experience-based approach has little to do with the naive vision of a harmonious society. Indeed the conflict among residents themselves is often more serious than the conflict between residents and the government or developers. In other words, it is the conflict among those who conduct their everyday life in the same area that is often more problematic. Governments (and political parties) and developers frequently exploit this conflict among residents. Given the importance of understanding the depth of mutual non-understanding among inhabitants, let us explore this issue in more detail. We now proceed to the second point.

Rejecting Harmonious [Social] Systems

In settings of everyday life (e.g., local communities such as towns and villages) where the environmental becomes a social issue, the opinions of inhabitants rarely align. First of all, each inhabitant's opinion is fundamentally unique, but everyday life requires some form of process to manage the diversity of opinions. Typically, individual opinions are eventually coordinated and sorted into two or three groups: those who approve, those who conditionally approve, and those who oppose. If these two or three groups remain divided and become socially visible, then they may be perceived by an observer as a division within local community organizations or movements. However, it quite often happens that some or all of these groups give in or reach a compromise, and the matter is settled before the division comes to the fore.

Either way, in the case of major environmental changes, it is highly unlikely for the diverse opinions of residents to converge and reach a solution that is acceptable to all. In reality, individual opinions tend to split into those several groups as described above. These grouped opinions are not identical to the opinions of specific individuals, nor do they seem to be a coordinated outcome of multiple opinions within each group. They manifest as qualitatively different opinions.

That is to say, whether it is the building of a community center, the redevelopment of a station-front area, the construction of a bypass road, or the construction of sewerage systems, it is normal for the opinions of inhabitants to be divided into groups. Accordingly, rather than formulating an analytical framework (a system of explanations) based on the assumption that people will eventually come to understand each other and achieve a harmonious system, we aim to develop a framework that recognizes the presence of fundamental non-understanding. In the field of sociology, the concept of life-world has become recognized in recent sociological theories, thanks to the social scientific works of Schutz on phenomenology. The study of everyday life by Berger and Luckmann (1966), which I understand to be a sociological interpretation of Schutz's work, offers useful suggestions. In particular, their study extensively incorporates the issue of "subjectivity" into the analysis of everyday life, which is intriguing.

Indeed, in the next section, I adopt the term "life-world" as a convenient analytical tool. However, regardless of my instrumental use of the term, I have fundamental doubts about the effectiveness of their theory itself. It feels as though the theory assumes an excessive degree of pre-established harmony and becomes detached from reality. In this chapter, I side with Naoharu Shimoda's commentary, despite it being a somewhat extreme characterization of Schutz's intellectual history as a whole. Shimoda states,

> What is noteworthy is the almost excessively conservative character of Schutz's theory. What an idyllic and peaceful world the "everyday world" is in his understanding! There is no indication of people, while sharing what Schutz calls common knowledge, engaging in conflicts of interests. There are no people absorbed in ideological battles based on conflicting worldviews, who may even resort to violence to the extent of killing others in some cases.
> (Shimoda 1981, 120–121 [original in Japanese])

I suspect that the reason why Schutz developed a theory of such an idyllic and peaceful world is that he focused solely on laying the foundation of the subjective world of everyday life within social science. Certainly, it was an important first step for theory building. If Schutz had further investigated into the question of the grounds for individuality, investigating whether and how the differences in people's subjective views originate, Shimoda's criticism might not have arisen. However, upon careful examination of Schutz's writings, we observe that he qualifies his theory of the idyllic and peaceful everyday life-world with phrases such as "in the natural attitude," "in general," and "other things being equal" (Schutz and Luckmann 1973: 3–20, 59). In other words, we could interpret that Schutz merely presented the theory of principles (and his original motivation, mutual understanding), recognizing that reality is not as idyllic and peaceful. However, it is also undeniable that most theorists influenced by Schutz's work, to varying degrees, have reinforced the argument focusing on the idyllic and peaceful worldview. Accordingly, as fieldworkers, while we appreciate the rich insights offered by the theory, we cannot help but insist on the fundamental detachment from reality inherent in its idyllic and peaceful orientation that presumes harmony.

State Power Permeating Everyday Life

The third point is the issue of power. This issue of power must be considered from the viewpoint of "morality," which is deeply ingrained in the consciousness of ordinary people. In this case, I use "morality" as one type of "life consciousness"[10] promoted by the state for the purpose of governing the nation. We can conceive of several different types of life consciousnesses that serve as criteria for actions in everyday life, including morality as defined here and those types that are formed through ordinary everyday experiences. Nevertheless, sociological theories about everyday life to date have been rather weak in considering the state-controlled morality. The reason for this is obvious. Take the edited volume by Lofland, for example, as a traditional sociological work on everyday life. It defines that the scope of sociological analysis as ranging from immediate face-to-face encounters through entire societies, and it focuses on the immediate human relationships within small areas in the analysis of everyday life (Lofland 1978: 6).

This is a very common way of defining everyday life in sociology. This definition leads to the view that the analysis of state power should be carried out in the analysis of entire societies, and can be excluded from the analysis of everyday life. However, even when analyzing face-to-face interactions in a small area, it is clear that people's actions are strongly influenced by the morality imposed by the state. What kind of analysis of everyday life would it be if this aspect is ignored?[11]

Life-environmentalism encompasses these three analytical characteristics described above. They are articulated based on reflections of our own past research, leading us to believe that conducting research with those methodological emphases in mind will allow for a more accurate understanding of the subject. Nonetheless, this kind of framework is something that should be continuously reviewed and revised throughout the research process. Just as this chapter is an outcome of revising my own previous work, I assure the reader that further revisions will be made. I would also like to note that the claims made in this chapter may not necessarily be accepted by all the authors of this book.

Everyday Experience and Decision-Making

Our World and Their World

The extent of people's everyday-life activities can be defined at various scales, such as family, neighborhood, village, suburb, and municipal area. When a social problem or concern arises, the researcher may determine the analytical scale of the shared life space based on the type of problem or concern. For instance, the family would be an appropriate scale for a married couple's problem, the neighborhood for an issue involving a garbage station, and the local community association (*jichikai*) for a problem of building a new assembly hall. Within each of these scales, inhabitants share their everyday life activities and life consciousness to a certain degree, depending on the nature of the problem or concern at hand. Let us call the conceptual space in which they share their everyday activities and consciousness the "life-world." It is important to note, nevertheless, that even when there is a certain degree of shared everyday activities and consciousness, individuals may still interpret and assign different meanings to a given problem or issue. In that case, we call our own interpretive and semantic world "our world" (*jikai*), while recognizing the world of individuals and groups that have different interpretations and assign different meanings, which we label as "their world" (*takai*). For example, these may take the forms of a group in favor of a particular issue ("our world") and a group against it ("their world").

When a researcher defines the categories of "our world" and "their world," it is common to use the attributes of group members as criteria for the definition. This is because attributes are easily observable markers for investigators. For instance, in the case of tenancy disputes before WWII, the definitions would be "we, landlords as the proponent group" and "they, tenant farmers as the opposition group." In the case of the reconstruction of the hall of *Kannon-ko*, a traditional gathering of married women to wish for easy delivery and good child-rearing, the definitions would be "we, women as the group supporting swift reconstruction" and "they, men as the group supporting the moratorium on reconstruction."

By the way, from the example of the male–female juxtaposition mentioned above, some readers may associate it with the well-known idea of binary opposition in the works of Lévi-Strauss and structuralism as an analogous concept. Indeed, Lévi-Strauss reveals "structure" by employing various binary oppositions, such as male/female and center/periphery. However, what I am doing here is not identifying opposing terms by examining males and females from a third-person perspective. Instead, I take the position where I stand on one side and view the other side, and vice versa. It is akin to seeing women from a male perspective and men from a female perspective. In other words, it is a way of thinking that divides the world into two: the world to which "we" belong and the world to which "they" belong.

It was the philosopher Junzo Karaki who pointed out that "Modern science and modern rationality render everything third-person" (Karaki 1949, 98, editors' translation). He made this claim in relation to religion and science, but it is worth revisiting today in a different context, four decades since Karaki first made the assertion. In short, this chapter aims to explore environmental problems within the two-persons world of "I and thou," to use the old expression. What I mean by the subjectivity of position is based on this understanding.

Three Types of Everyday Knowledge

If we assume that each individual or group has their own world (i.e., "our world"), how does it relate to their actual actions? As mentioned earlier, the shared everyday-life consciousness among people forms the life-world, which is rooted in each person's experiences. This is because one's own experiences form one's life consciousness, which, in turn, shapes the life-world. This life consciousness rooted in experience is important because when one engages in concrete actions, it serves as knowledge informing the decision. Let us call this knowledge "everyday knowledge." Then, the everyday knowledge serving as the criteria for determining an individual's actions can be divided into the following three categories.

(a) Personal experiential knowledge (*taiken-chi*; not experience itself, but knowledge formed as the result of one's experience)
(b) Living common sense (*seikatsu jōshiki*) within life as organizations (such as traditional village communities, *mura*)
(c) Popular morality (*tsuzoku dotoku*) brought in from outside life organizations

Personal experiential knowledge (a) refers to knowledge acquired through one's own personal experience. In contrast, (b) and (c) are shared norms acquired by individuals from the outside. While there can be various types of knowledge from external sources, these two are probably the most important types of externally originated everyday knowledge that influences individuals' actions in the context of environmental problems.

When we look at the span of 100 years of modern Japan, it is remarkable to see the strong influence of what is commonly known as national morality, which the Japanese state has exerted over the nation. Before World War II, national morality was succinctly expressed in the Imperial Rescript on Education (1890). However, the actual national morality embraced by the nation was not so much the complex reasoning presented in the Rescript, but rather virtues such as diligence, frugality, filial piety, and honesty, which formed its foundation. It is worth noting that these virtues were not unique to modern times but can be traced back to the early modern period. What remains consistent is that they served ideological functions in governing the nation under state control.

Historian Yoshio Yasumaru defined this phenomenon as "popular morality" (*tsūzoku dōtoku*), and described it as "nothing but the ideological superstructure of Japanese capitalism that unfolded rapidly while being complemented by various feudal relations" (Yasumaru, 1965, 1–2; editors' translation). While one may have different opinions about this specific definition from the Kōza school,[12] the fact remains that national morality served as an ideology for governing the nation. Therefore, this paper follows in the footsteps of Yasumaru and refers to it as "popular morality."

Yasumaru (1965) pointed out that popular morality was "the most important mechanism for dealing with various challenges and contradictions (such as poverty) in modern Japanese society" (1, editors' translation). He says,

> If I am poor, popular morality teaches me that I am not diligent enough, and if my family is in discord, it teaches that I lack filial piety, and so on. Consequently, various challenges and contradictions are mistakenly attributed to my everyday attitude and practical ethics, and they are processed within this illusory assumption.... In the meantime, professional advocates and devotees of this illusion arise, diverting people's attention from recognizing various real relations. They persuade individuals to perceive everything real through this illusion, thereby creating a system of ideological domination.
>
> (2, editors' translation)

For ordinary individuals, apart from certain intellectuals, a moral framework has been established where living a proper life as a human being means conforming to popular morality. When faced with societal contradictions, people tend to believe that their unhappiness is not caused by the problem of the society but rather due to their own unworthiness.

At the site where social problems arise, it is undeniable that popular morality serves as a criterion for individuals to make decisions and also as a logical basis for persuading others. When people embrace popular morality, they effectively position themselves as actors within the governing ideology. However, since governance operates indirectly, people are unaware that they are actively taking part in national governance. This is a characteristic feature of popular morality.

Based on my field research experience, it is the Japanese emigrants [editors' note: during the pre-WWII period] who strongly adhere to popular morality among other externally originated norms (i.e., knowledge). They typically emigrated after completing elementary school at the age of 14. Their learned norms primarily consisted of virtues instilled by their elementary school teachers. It goes without saying that these virtues constitute popular morality.[13]

On the other hand, ordinary farmers and fishermen, after completing elementary school at the age of 14, would receive a "coming-of-age education" from the village community to become independent farmers and fishermen. A village (*mura/buraku*) is a local life organization with well-defined spatial boundaries. Traditionally, Japanese villages had maintained a certain degree of economic, political, and social autonomy, embodying the characteristics of a management body. To sustain the functioning of the village, it was essential to retain self-reliant individuals/workers, often referred to as "*ichinin-mae*" [editors' note: roughly translated as "full-fledged person"] or other equivalent terms specific to regions. This was the purpose of the "coming-of-age education." In other words, the village also functioned as an educational institution.

What kind of education was provided there? In addition to teaching production techniques, youths were also taught how to navigate the world as an average person; in other words, they were taught common sense. This education covered a wide range of topics, including how to interact with individuals of different attributes such as elders, leaders, and people of the opposite sex, how to get along with fellow villagers, how to show respect to local deities and mountain gods, and how to fulfill assigned tasks and roles. We can refer to this wisdom, commonly found in villages, that guides a harmonious everyday life as "living common sense." By conceptually defining this "living common sense" as the accumulation of wisdom for a better everyday life within one's own community, the distinction from "popular morality" should become clear.

Sociologist Kizaemon Aruga's concept of "life consciousness (*seikatsu ishiki*)" and Eitarō Suzuki's idea of "spirit of a natural village (*shizenson no seishin*)" are closely related to the concept of "living common sense" discussed here.[14] In addition, it is worth highlighting the insights of historian Daikichi Irokawa, who clearly recognized the distinction between "living common sense" and "popular morality":

> I would like to address the relationship between the morals of the village community (author's note: meaning "living common sense") and "popular morality." In my previous papers, I have not adequately clarified the relationship between the two concepts, and sometimes mistakenly treated them as interchangeable. In reality, the *buraku* community (author's note: meaning "village") has its own distinct principles of living and self-contained morals that have developed over many years. These principles become evident in times of crisis through practices such as a "village ostracism." During normal times, they are observable during village community events, or as a form of social rules to foster the development of *kodomo-gumi* (translator's note: community-based children's group) and *wakamono-gumi* (translator's note: community-based young adult group). I had previously failed to differentiate these internally derived morals of the village community and exogenous popular morality, nor had I examined the dynamic interaction between the two.
>
> (Irokawa 1974, 257, editors' translation)

Irokawa emphasizes the significance of the concept of "living common sense" by associating it with the village (*mura*). However, it is important to note that "living common sense" is not exclusive to rural village organizations. It can also be found in urban regional organizations, albeit in different forms than those typically observed in villages. Additionally, scholars such as Yoshirō Tamanoi, who specialize in regionalism, advocate for the importance of "living common sense" and argue for the establishment of such regional organizations in areas where they do not currently exist (Tamanoi 1977, 1978).

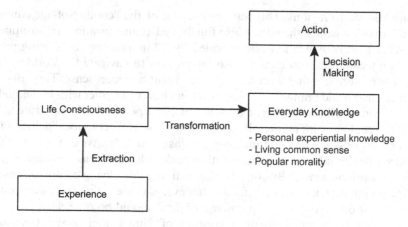

Figure 18.1 Process of forming a specific action based on experience.

Above I have discussed the distinctions among the three types of "everyday knowledge": (a) personal experiential knowledge, (b) living common sense, and (c) popular morality. In everyday life, individuals make decisions based primarily on these three types of everyday knowledge (which are acquired through their "experience"), and these decisions become social actions (See Figure 18.1).

With this understanding of everyday knowledge, let us now proceed to the next stage: decision-making.

The Patterns of Decision-Making

Table 18.1 shows the patterns of decision-making in the local community, considering the aforementioned understanding of everyday knowledge. The right side of the table indicates the relationship with everyday knowledge.

- Pattern (1): "Listen to the opinions of others, but it is permissible to disregard those opinions." A good example of this can be seen in the fishing village of Koza, Wakayama Prefecture. Even when fishermen fish individually, they get together on the beach and freely discuss various topics, such as the weather, based on their own experiences. Each person listens to others' opinions, agrees or disagrees, but ultimately makes their own intuitive decision to go or cancel fishing. Consequently, each fisherman takes responsibility for their catch or any accidents that may occur due to rough seas.
- Pattern (2): "After each person expresses their opinions, the leader makes a judgement that goes beyond reason." In the case of fishing in groups of Koza, the *oyakata*, who is the head of the group, silently listens to each person's opinion, and then declares, "We will go fishing today," or "we will not." Group members must obey the group leader's decision. Because the

Table 18.1 Patterns of Decision-Making in Local Communities

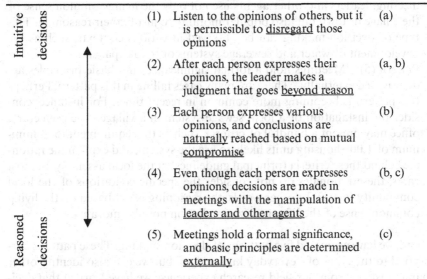

Intuitive decisions ↕ Reasoned decisions	(1) Listen to the opinions of others, but it is permissible to <u>disregard</u> those opinions	(a)
	(2) After each person expresses their opinions, the leader makes a judgment that goes <u>beyond reason</u>	(a, b)
	(3) Each person expresses various opinions, and conclusions are <u>naturally</u> reached based on mutual <u>compromise</u>	(b)
	(4) Even though each person expresses opinions, decisions are made in meetings with the manipulation of <u>leaders and other agents</u>	(b, c)
	(5) Meetings hold a formal significance, and basic principles are determined <u>externally</u>	(c)

Types of "everyday knowledge"
(a) *Living common sense* within life organizations (such as *mura*)
(b) *Living common sense* within life organizations (such as *mura*)
(c) *Popular morality* brought in from outside life organizations

leader is selected based on his wealth of experience, the members generally do not demand an explanation for his decisions, trusting his experience. The fishing group is an example of a village life organization.
- Pattern (3): "Each person expresses various opinions, and conclusions are naturally reached based on mutual compromise." Let us consider the case of Tsushima-Ina Village. During village meetings, village members come together to discuss various issue in the village, sometimes splitting into smaller neighborhood groups, to reach conclusions. If a tentative conclusion is reached in a neighborhood group, it is brought to the village head. If they are unable to come to an agreement, they return to their own neighborhood group for further discussion. These discussions may last more than a day, but participants are free to leave if they have other commitments or feel tired. At first glance, these meetings may look like aimless chats, with diverse topics and scattered discussion. However, as people bring up precedents and engage in small talk, they gradually come closer to a conclusion. Through the exchange of various stories, natural conclusions are reached for each agenda item (Miyamoto 1971, 711). In this case, villagers are likely to be satisfied when judgments are made in accordance with the common sense of the village.
- Pattern (4): "Even though each person expresses opinions, decisions are made in meetings with the manipulation of leaders and other agents." This

is the most common form of decision-making. To persuade attendees of the meeting, leaders and other agents use not only the living common sense of the village or town but also popular morality to justify their reasoning. This type of decision-making can be observed in various contexts, such as the development of water and sewerage systems or road expansions.

- Pattern (5): "Meetings hold a formal significance, and basic principles are determined externally." There are many cases falling in this pattern. Perhaps this pattern is becoming more common in recent times. For instance, consider the installation of a local sewage system in a village. The prefectural office may present a handful of criteria, such as the requirement for a minimum of 1,000 housing units likely to use the system, and explain the rationale behind these criteria (principal guidelines) at the local assembly. In most cases, these rationales do not consider the specific conditions of the local community (*mura*). In other words, the reasoning is not based on the living common sense of the life organization but on popular morality.

Above, we have examined five patterns of decision-making. These patterns correspond to the types of "everyday knowledge," but we can also identify other characteristics. From our field research experience, we have learned that decisions are often derived not only through logical reasoning but also through intuition. Based on our classification, Pattern (1) leans more toward intuition-based, while Pattern (5) represents the most reasoned decision, as indicated by the arrow on the left side of the table. It is important to note that the distinction between the intuitive and the reasoned is a matter of degree. I am not suggesting that Pattern (1) completely lacks reasoning entirely, but rather emphasizing the substantial role played by intuition. Nonetheless, I believe it is necessary to demonstrate, through this classification of decision-making patterns, that decisions are not solely based on reasoned judgment.

Actual Methods of Problem-Solving and Life Organizations

Presence of "Back Patterns"

In the previous section, we discussed five patterns of decision-making. While these five patterns may initially appear sufficient to explain the actual decision-making process, the reality is not so simple. This raises the question of whether we can accurately describe actual decision-making by increasing the number of patterns to seven or even twenty. While a greater number of patterns would provide broader coverage, it would also reduce our ability to logically interpret the problem. In that sense, a simplified scheme is more practical for real-world applications.

What I say, "[the reality] is not so simple," I mean that there are not as many real-world cases as one might expect that strictly follow these five patterns of decision-making. There are in fact many instances that go against each of these five patterns. If we consider the five patterns in Table 18.1 the "front" patterns, there are also "back" patterns ("back" in this case is not used in a negative sense). These are:

1 When a person who listens too much to the opinions of others is unable to make their own decisions, relying on others for decisions.
2 When each person expresses their opinion and the leader summarizes them into multiple opinions and processes them.
3 When each person expresses various opinions and the progress of mutual compromise halts at a certain stage.
4 When the manipulation of the leader or other agents fails, and each person's opinion begins to take on a life of its own.
5 When the external guidelines/principles themselves begin to be questioned at the meeting.

Even when one of the back patterns occurs, in most cases, a decision is eventually made and the matter is settled. However, the difference from the front patterns is that there are conflicts of opinions that can be sorted into several groups.

From another perspective, this back pattern may well be the norm in the case of social problems such as environmental conflicts; the front pattern can be seen as lucky exceptions. At the very least I regard the front patterns, as shown in Table 18.1, as logical intermediaries for us to properly grasp the back patterns. That is, because it is not easy to articulate the nature of the back patterns, I first showed the front patterns in order to present the back patterns as their mirror constructs. The significance of the front patterns should be understood as such.

Understanding that the back patterns are actually the norm, meaning that there are groups of people representing different opinions, let us consider how this situation relates to life organizations.

Environmental Alteration and Life Organization

Figure 18.2 shows the process through which environmental changes impact local community life.[15] As mentioned before, the scope of life activities can vary depending on the specific focus of analysis. In this case, we will concentrate on the *mura* (*buraku*) as the primary extent of life activities and examine the process in detail. The demand for environmental alteration will be felt by the villagers as an imminent impact to foster changes in existing living conditions or to impose entirely new living conditions. Consequently, they need to find a way to manage these impending changes in their living conditions. The filter through which they manage and process these changes is the "everyday knowledge" discussed earlier. This filter allows for some adjustments to be made to the new living conditions. While there are instances when the new conditions are accepted without modification, it is more common for the villagers to incorporate their own strategies and modify the conditions to align with their everyday

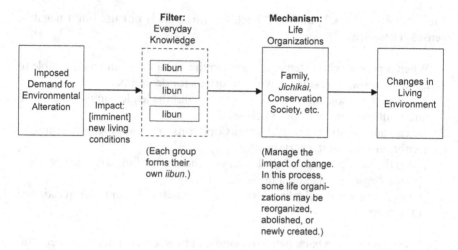

Figure 18.2 Process leading to changes in living environment.

life. This is because the new living conditions do not fit into their everyday life without some modifications. The permissible extent of these modifications, including the possibility of rejecting the new conditions entirely or not making any changes at all, will vary depending on each individual's standpoint.

It goes without saying that the challenge lies within formulating these strategies. Even the village headmen during the Edo period could not decide on these strategies single-handedly. In the present day, with the multitude of diverse opinions and arguments reflecting different standpoints, the choice of strategy becomes a crucial issue.

Upon a close look these standpoints vary significantly from one person to another. As individuals constitute the smallest unit of living, it is only natural for each person to have their own unique life-world. However, in reality, groups tend to form based on the degree of expected modification in their living conditions. When the issue is recognized as a clear social problem among the residents, it is common to observe groups such as the "approval group," "conditional approval group," and "opposition group."

Individuals with their own unique life-world begin to interact with others regarding the issue at hand. These interactions occur based on attributes related to the issue; those may be landowner/tenant, direct-victims/indirect-victims/non-victims, men/women, old/young, villagers/suburbanites/urbanites, and so on. They form shared life-worlds referred to as "our worlds" and start engaging with groups that possess different attributes, known as "their worlds." For instance, if we consider the group of landowners as "our world" when viewing the focal issue from their standpoint, then the group of tenant farmers would be considered "their world." Through this process of interacting with the "their world," "our world" is reshaped, and individual opinions are transformed, becoming more tangible, leading to the formation of groups that foster mutual recognition of shared perspectives.[16] The formation of these groups and the underlying logic are crucial and warrant further examination, which will be presented in the following section, without scrutinizing the details here.

Each of these groups formed as the result of these dynamics develop their own *iibun* (logic of justification). Each group's *iibun* is based on "everyday knowledge." Consequently, within the tension of power relations among the groups, with each group supporting their own *iibun*, the social problem at hand is processed and resolved.

The mechanism used to process the problem is the "life organization."[17] In a village, there are multiple such mechanisms, and the type of problem determines which life organization is employed. For example, for the issue of waste management, the local *jichikai* (neighborhood association) is often utilized as a suitable life organization. This life organization processes the impact on living conditions, which is modified by "everyday knowledge," or, more specifically, by the logic (i.e., strategy) ultimately determined by the power relations of the groups supporting their own *iibuns*. During this time, the life organization is usually reorganized to varying extents. That is, the life organization itself is not constant, but continues to transform. In addition, if the reorganization of the life organization is not sufficient, it may be abolished or a new one may be created. For example, if the waste management problem is too large for the local *jichikai* to handle, a new organization such as the "Group of Those Who Are Concerned About Waste Management in District X" may be established.[18]

In these ways the life organization functions as a mechanism to respond to and manage impacts while undergoing its own transformation. As a result, (local) everyday life itself will undergo changes as well. This is the process that starts from an imposed demand for environmental alteration and culminates in the eventual transformation of in (local) everyday life.

The Formation of Group Iibun

Let us now turn out attention to the critical issue of *"iibun"* (logic of justification), which I have yet to fully elucidate.

To that end, I believe it would be helpful to present a somewhat exaggerated thesis, "We cannot understand the mind of a person, but we can understand the mind of a group of people."[19] There are various types of fieldworkers, and I consider myself to be of the tenacious type. I have never finished my fieldwork in a short period of time. I am among those who spend several years, or even over a decade, immersed in the field. Even when I use a questionnaire survey, I habitually include open-answer sections, and am never satisfied with statistical analysis alone. This type of research allows me to gather detailed information, which is an advantage, but there is a negative side as well—a limited number of study sites often leads to a narrow perspective. In any case, it is a tenacious type of investigation. Even then, the study of individual "consciousness" (or, ways of thinking, broadly speaking) remains a significant gap in my research, and I am acutely aware of this. I have conducted research on the life histories of 30 to 40 individuals, yet even after learning their entire life histories, it remains exceedingly challenging to articulate the nature of their consciousness. Sociological research methods appear to have an inherent limit where we cannot gain deeper insights into the consciousness of each individual human being. Upon contemplation, though, it may not be solely a matter of research methods; even the individual in focus may lack a clear understanding of their own consciousness. This is what I mean by "We cannot understand the mind of a person."

Regarding "the mind of a group of people," I am not referring to a mere shared consciousness among multiple human beings. Instead, it refers to "*iibun*" (logic of justification) as defined in this chapter.

Each person has their own consciousness, and it is not necessary to inquire about how clearly each individual is aware of it. In any case, when a certain problem or issue arises, those who possess their own unique consciousness will collaborate with others who share the common "our worlds." Those groups may be formed among peasants, among women, among young people, among longtime locals, and so on. The existence of "our world" is contingent upon the existence of "their world," and "our world" maintains its distinct identity by engaging in ongoing interaction with "their world." For example, landowners represent "their world" for peasants, men for women, newcomers for longtime locals, townspeople, or city dwellers for rural villagers, and so forth. They interact with "their world." It is important to note that this "interaction" does not necessarily mean getting along well with each other. Rather, it means to be clearly aware of the existence of "their world," which helps "our world" to recognize its own uniqueness.

I suspect that the formation of "our world" occurs consciously only when a specific problem arises. When those who comprise "our world" seek to address the issue at hand, they develop a certain logic to convince themselves and persuade others. This logic is what I call "*iibun*." This *iibun* is more or less logical because its goal is to convince themselves and persuade others, and it is grounded in the "everyday knowledge" discussed earlier. Other kinds of knowledges are generally not effective in persuading individuals within "our world" and those outside of it. Particularly, in the process of persuasion, (b) living common sense and (c) popular morality are likely emphasized. As I have already mentioned in the explanation of Table 18.1, these two types of "everyday knowledge" exhibit a more logical nature than (a) personal experiential knowledge.

Since each group forms an "*iibun*" based on the "everyday knowledge" [within their community], resulting "*iibuns*" often exhibit some similarities. Sometimes these groups explain to researchers that a critical essence resides in those subtle differences. However, my understanding is that the problem lies in "our worlds" that support the *iibuns*. Each "our world" has an indisputably unique quality that distinguishes itself from others, and there is no such thing as a subtle degree of difference from others. It is only that the emergent "*iibuns*" sometimes happen to show subtle differences.

What matters here, though, is that people begin to act according to the logic of the already established "*iibun*," and begin to debate over the legitimacy of their group's "*iibun*," expressing it as their own opinion. The "mind of a group of people" in this chapter refers to these opinions that are held by multiple individuals and are asserted by the members of the group as their personal opinion. The "mind of a group of people" can be grasped by sociological research because it is a form of logic so that research respondents can also logically explain it as their own opinion to the researcher (in this case, the word "logical" is used in a very broad sense, meaning "to explain with reason"). This is why we, as researchers, can also comprehend it when we listen to their explanations.

To reiterate, people's actions are primarily driven by *"iibun"* rather than their individual consciousness. This is why understanding *"iibun"* becomes the most important issue for researchers.[20] However, it is crucial to descend to and study the "experience" of each individual and organization in order to understand the essence of an *"iibun"* and the direction of its change. Life history and historical analysis become essential tools for this purpose.

In conclusion, I have presented a framework for approaching environmental problems within the context of everyday life. Research frameworks can be constructed in various ways, but the underlying intention of this chapter has been to provide an "empirically enabling framework." For this reason, the framework presented here may seem as somewhat austere to some readers. For instance, if we incorporate the concept of "intersubjectivity" from phenomenology, we may be able to better articulate the realm of non-logic (or the world of intuition). However, while this concept may be theoretically valid, I feel it is somewhat unsuitable for empirical research. In other words, considering the sociological research methods currently available, I believe that the framework presented here is at the most practical level of approach.

Notes

1 Don't overanalyze the usage of the word "grammar" in this context. While there have been recent discussions in the social sciences regarding language and *parole*, influenced by linguistics, the term "grammar" here does not carry such specialized meanings. It simply refers to grammar as commonly understood when learning a foreign language.
2 I think that the theory discussed in this chapter closely aligns with the following definition by the statistician Chikio Hayashi, so I would like to introduce it (emphasis added):

> When dealing with social phenomena, a theory that falls under the category of "theory is now proven" would be something like "if it rains, the weather is bad" or "if the dog's head points east, then the tail points west." However, theories that do not fall into this category are not strictly meant "to be proven." Instead, *they should be evaluated based on whether the various inferred corollaries from the theory are useful in efficiently discovering the predicted "relationship between data."* We need to generate a theory based on the relationship between data; it is not that a theory is proven by data itself. The theory should be refined through repeated analogical inferences and the acquisition of wisdoms based on them. The validity of the theory is determined by its ability to effectively suggest the upcoming "relationship between data." Empirical proof and scientific utility lie within the "relationship between data." By thinking in this manner, the "theory for the elucidation of social phenomena" should provide an opportunity to better predict the forthcoming relationship between data.
>
> (Hayashi, 1974, 104–105, editors' translation)

3 Editors' note: In the original Japanese text the word *"shugi"* [-ism, principle, ideology] is used here. As explained below, the use of "-ism" is intended to express that life-environmentalism is a particular ideological position.
4 The content of the Stockholm conference is directly reflected in the cited report. However, a comprehensive understanding is made difficult by the report's extensive length and wide-ranging scope. In contrast, the report *Beiträge zur Umweltgestaltung*, distributed by West Germany in 1982 during a session of the United Nations Environment Programme, effectively summarizes the key points of the conference. It conveniently includes a table that demonstrates the levels of implementation of the action plan.
5 Thinking along the lines of "sustainable development" is not necessarily new. However, it was *World Conservation Strategy* published by the International Union for Conservation of Nature and Natural Resources (1980) that first directly expressed the idea. The subtitle of this report is "Living Resource Conservation for Sustainable Development."
6 Editors' note: In Japan the term "modern economics" (*Kindai Keizaigaku*) refers to both neoclassical economics and Keynesian economics, or to what is not Marxian economics.
7 The movement based on this principle is commonly called the ecology movement. This movement has a history of nearly a century of struggle, leading to the now widely accepted notion of "harmony between nature and the environment." We regard the history of such movements as valuable (for example, Tadashi Yoshida's *Seitaigaku to Ekorojī* (Ecological Science and Ecology) (1984) serves as a useful reference for understanding the distinction between ecology as an academic discipline and *ekorojī*). However, we believe that life-environmentalism that views "humans in relation to nature," rather than natural environmentalism that views "humans as part of nature," has a more practical advantage in the actual fields of environmental problems. From our standpoint, for example, we would choose to conditionally support whaling by Japanese fishermen, despite criticisms from nature conservation activists in other countries.
8 The term was first used with this nuance by Kazuko Tomiyama (1980, 180).
9 Editors' note: The original term 相互無理解 (*sōgo murikai*; *"mutual non-understanding"*) is not a commonly used Japanese word, unlike the term 相互理解 (*sōgo rikai*; *"mutual understanding"*). One might be tempted to translate the term to more familiar English words such as "disagreement" or "misunderstanding." However, we believe that *sōgo murikai* has an important nuance of "both sides are thoroughly unable to comprehend each other even when they try," rather than actively opposing the other person (disagreement) or lacking sufficient communication (misunderstanding). Therefore, in this book, we use the literal translation of "mutual non-understanding" while acknowledging the awkwardness of the term.
10 In the next section, this "*seikatsu ishiki*" (life consciousness) is rephrased as "*nichijōteki na chishiki*" (everyday knowledge). This is because "life consciousness" is also used to refer to concrete knowledge in everyday life.
11 H. Lefebvre's exposition is one of the few examples to connect everyday life and the social complex. He develops his argument around the question of what should be done for the alienated workers to regain life in the true sense of the term. He writes,

> "The relation of every humble, everyday gesture to the social complex, like the relation of each individual to the whole, cannot be compared to that of the part to the sum total or of the element to a 'synthesis', using the term in its usual vague sense. Mathematical integration would be a better way of explaining the transfer from one scale of greatness to another, implying as it does a qualitative leap without the sense that the 'differential' element (the gesture, the individual) and the totality are radically heterogeneous. Within the parameters of private

property, this relation of the 'differential' element to the whole is both disguised and distorted. In fact, the worker works for the social whole; his activity is a part of 'social labour' and contributes to the historical heritage of the society (nation) to which he belongs. But he does not know it. He thinks he is working 'for the boss'.... Integration takes place beyond the will of individuals, outside of their 'private' consciousness".

(Lefebvre 1991[1947], 164)

This approach of linking everyday life to the social whole is commendable, but this perspective, rooted in the alienation of labor, is too general to be of direct utility for the purpose of this chapter.

12 Editors' note: The Kōza school was a group of Japanese Marxist scholars formed in the 1930s.
13 For a specific example of how the popular morality of the Imperial Rescript on Education had a strong impact on the way of living of the common people, the "Ōinaru Shōjiki (Great Honesty)" chapter in Hiroyuki Torigoe's *Okinawa Hawaii Imin Issei no Kiroku* (The Record of the First-Generation Immigrants to Okinawa and Hawaii) (Chuko Shinsho, 1988) will be helpful.
14 For introductory explanations of the concepts of "life consciousness" and of "spirit of a natural village," refer to *Ie to Mura no Shakaigaku* (Sociology of Ie and Mura) by Torigoe (Sekai Shisōsha 1985, 77–82).
15 Based on Torigoe (1983, 162) with some revisions.
16 Attribute themselves can also change at this point. As highlighted by Motoji Matsuda, my co-investigator in the Lake Biwa research, attributes should not be regarded as "fixed and immutable entities given *a priori*." Also, these attributes are in part a collective illusion by the individuals involved (Matsuda 1985, 45–46).
17 The concept of "life organization" was originally defined by Kizaemon Aruga. See Torigoe (1982, 383–384) for detailed explanations of this concept.
18 Note that I use the wording "the logic (i.e., strategy) ultimately determined by the power relations of the groups supporting their own *iibuns*," rather than "... determined through dialogue of the groups supporting their own *iibuns*." In reality, resolutions are determined by power relations rather than by mutually learning beneficial points from each other through dialogue. Such a conversation is not feasible because each group operates with different language games (rules). When one group communicates with another group, the latter group interprets it according to their own set of rules. Therefore, even if two parties have the opportunity to converse, they are effectively speaking past each other rather than engaging in true understanding. The only ones who can truly comprehend are individuals within the same group (who share the same language game).
19 This expression is borrowed from Hiroshi Nagasaki (1977, 186). However, it is being used here in a very different context.
20 Let's clarify the meaning of *"iibun"* once again. *"Iibun"* consists of three essential conditions. First, it is holistic. It is often emphasized that an individual's actions encompass the whole. For instance, Goerner uses de Gaulle as an example. When de Gaulle states in his autobiography that he was motivated by his idea of France, this image of France is holistic (Gellner, 1968, 259). Second, it asserts justification. Justification is based on the understanding that not everyone appreciates a particular claim or movement. Thus, it becomes necessary to persuade others through justification, as discussed in the main text, primarily relying on living common sense and popular morality. Strictly speaking, this justification often involves makeshift explanations (defensive excuses and ex post reasoning). Third, it is accompanied by "reason." In many cases, these "reasons" are derived from past experiences known to the group. For example, if a neighboring district had a failed prefectural pilot project for sewage purification, it may lead to the reasoning of "consequently, let's reject the prefecture's pilot project in our district."

References

Allport, Gordon W. 1962. "The General and the Unique in Psychological Science." *Journal of Personality*, 30 (3): 405–422.
Berger, Peter L. and Thomas Luckmann. 1966. *The Social Construction of Reality*. New York: Doubleday.
Beiträge zur Umweltgestaltung. 1982. Berlin: Erich Schmidt Verlag.
Documents for the U.N. Conference on the Human Environment. 1972, June 5–16. Stockholm, Sweden.
Gellner, E. 1968. "Holism versus Individualism." In *Readings in the Philosophy of the Social Sciences*, edited by May Brodbeck, 254–68. New York: The Macmillan Co.
Hanson, Norwood R. 1971. *Observation and Explanation: A Guide to Philosophy of Science*. New York: Harper & Row.
Hayashi, Chikio. 1984. *Sūryōka no Hōhō* [Methods of Quantification]. Tokyo: Tōyō Keizai Shinpōsha.
International Union for Conservation of Nature and Natural Resources (IUCN). 1980. *World Conservation Strategy*. Switzerland: IUCN.
Irokawa, Daikichi. 1974. "Kindai Nihon no Kyōdōtai" [The Modern Japanese Community]. In *Shisō no Bōken* [Adventure of Thought], edited by Kazuko Tsurumi and Saburō Ichii. Tokyo: Chikuma Shobō.
Karaki, Junzō. 1949, new ed. 1963. *Gendaishi e no Kokoromi* [An Attempt at Modern History]. Tokyo: Chikuma Shobō.
Lofland, John. 1978. *Interaction in Everyday Life*. Beverly Hills: Sage Publications.
Lefebvre, Henri. 1991 [1947]. *Critique of Everyday Life*. Translated by John Moore. New York: Verso. (Japanese translation published in 1968 by Hitohiko Tanaka. Tokyo: Gendai Shisōsha).
Matsuda, Motoji. 1985. "Afurika Toshi ni Okeru Dentō no Hirenzokusei ni Tsuite" [The Discontinuity and Continuity of Tradition]. *Jinbun Kenkyu* [Studies in the Humanities] 37(2): 79–112.
Miyamoto, Tsuneichi. 1971. "Tsushima nite" [At Tsushima]. In *Miyamoto Tsuneichi Chosakushū* [Works by Miyamoto Tsuneichi] vol 10. Tokyo: Miraisha.
Murakami, Yōichirō. 1979. *Kagaku to Nichijōsei no Bunmyaku* [The Context of Science and Daily Life]. Tokyo: Kaimeisha.
Nagasaki, Hiroshi. 1977. *Seiji no Genshōgaku arui wa Ajitētā no Henrekishi* [The Phenomenology of Politics or the History of the Agitator]. Tokyo: Tabata Shoten.
Popper, Karl. 1959. *The Logic of Scientific Discovery*. New York: Basic Books.
Schutz, Alfred and Thomas Luckmann. 1973. *The Structures of the Life-World*. Evanston, Illinois: Northwestern University Press.
Shimoda, Naoharu. 1981. *Shakaigakuteki Shikō no Kiso* [Fundamentals of Sociological Thought]. Tokyo: Shinsensha.
Tamanoi, Yoshirō. 1977. *Chiiki Bunken no Shisō* [The Idea of Regional Decentralization]. Tokyo: Tōyō Keizai Shinpōsha.
Tamanoi, Yoshirō ed. 1978. *Chiiki Shugi* [Regionalism]. Tokyo: Gakuyō Shobō.
Tomiyama, Kazuko. 1980. *Mizu no Bunkashi* [Cultrual History of Water]. Tokyo: Bungeishunjū.
Torigoe, Hiroyuki. 1982. *Research on the Tokara Island Society*. Tokyo: Ochanomizu Shobō.
―――. 1983. "Chiiki Seikatsu no Saihen to Saisei" [Reorganization and Revitalization of Local Life]. In *Chiiki Seikatsu no Shakaigaku* [Sociology of Community Life], edited by Michiharu Matsumoto. Tokyo: Sekai Shisō Sha.

Watanabe, Itaru. 1976. *Ningen no Shūen* [The End of Humans]. Tokyo: Asahi Shuppansha.
Weber, Max. 1972 [1922]. *Wirtschaft und Gesellschaft*. Tübingen, Germany: J.C.B. Mohr. Translated as *Shakaigaku no Konpon Gainen* (1972) by Ikutarō Shimizu. Tokyo: Iwanami Shoten.
World Commission on Environment and Development. 1987. *Our Common Future*. Oxford: Oxford University Press. Translated as *Chikyū no Mirai o Mamoru Tameni* (1987) by Saburō Ōkita. Tokyo: Fukutake Shoten.
Yasumaru, Yoshio. 1965. "Nihon no Kindaika to Minshū Shisō (Jō)" [Modernization of Japan and Popular Ethics (Part 1)]. *Journal of Japanese History* 78, 1–19.
Yoshida, Tadashi. 1984. "Seitaigaku to Ekorojī" [Ecological Study and Ecology]. *Rekishi to Shakai* [History and Society] 4: 105–132.

Index

Pages in *italics* refer to figures, pages in **bold** refer to tables, and pages followed by "n" refer to notes.

Act on New Industrial Cities (1962) 77
action-based research (*kōi-ron*) 32, 261, 263–4
actor network theory 29–30
adaptive governance 249n8
Air Dose Committee 92, 99, 102–3
Allport, G.W. 264
ambiguous loss 156–7, 160–1, 164
apparitions 156–7, 164; and Buddhism 161–3; and taxi drivers 158
Aruga, Kizaemon 28, 225, 271, 281n17
Asia xv, xvi–xvii, 1, 4–7, 10, 15–7, 35, 75, 86, 169, 184, 191–2, 257
Asian fawn lilies (*Erythronium japonicum*) 238
Asian Financial Crisis in 1997 184

Bailishihui (White Council) 111–2, 114
Benedict, Ruth 7–9
bengi-chi see expedient knowledge
Bon (Festival) 170, *171*–6
Boss, Pauline 156–7
Buddhahood 160–2
Bukseongro (Daegu City) 182–4, 188–9; Historical Architecture Restoration Project 187; tool museum *188*

Canada 259
capitalism 28, 59, 269
Carson, Rachael 208
chaxugeju 107–8, 115
Chen, Kuan-Hsing 10, 17, 75–6
China 1, 4–6, 15, 45, 105, 217
Citizen's Solidarity for Ecological Street (CSES) 184–5
civil society 1–2, 10, 15–6, 17, 31, 39n9, 75–6, 98, 120

Clean Air Act 1970 (USA) 208
Clean Water Act 1972 (USA) 208
climate change 1, 28
colonialism 5–6, 60, 181–4, 188–9, 209, 214–5
Committee on Harie Water Village 51
commons: dispossession of 149–53; forest 143–4, 146–9, 153; space 118–20, 217; tragedy of 118–9; two types of 126; water 49–50, 55
community: different definitions of 30–1
Conference on Common Property Resource Management 119
COVID-19 46
Cultural Revolution 110, 112

Daegu City 182, *183*; Rediscover Daegu Citizen's Movement 184–6
deep ecology 10–1, *13*–*4*, 15, 21n3, 24, 244
democracy 6, 8, 16, 21n2, 74, 185
democratic: consensus-building 58–9; consensus formation 72, 214; deficit 17
developmental: project 74–5, 149; state 6, 35, 74, 153

Earth Summit (1992) 239
ecological modernization 10–1, *13*–*4*, 15, 25
Edo period 28, 39n7, 85, 144, 149, 158, 172, 276
emotional sensitivity 32, 33–4, 228, 243, 247, 260–1, 263–5
environmental justice 1–3, 10–2, *13*–*4*, 16–7, 208, 215, 241, 243–4, 249n6
environmental pragmatism 236–7

environmental racism 211–2
Europe 4–5, 7, 10–1, 15, 19, 28, 32, 209
evacuation 92, 96, 103n2, 132, 142, *143*
everyday knowledge 75, 76–7, 217, 240, 242, 269–72, **273**, 275–8, 280n10; expedient aspect of *see* expedient knowledge
Evidence-based Policy Making (EBPM) 248
expedient knowledge (*bengi-chi*) 74–5, 77, 84, 86n2, 217, 225, 242, 249n5
experience (*keiken*) 55, 59–60, 76, 95, 101, 103, 143, 146, 153, 157, 163, 181, 185, 188–9, 209, 231–2, 246–7; and action (*kōi*) 32–3, 227, 242–3, 261, 263–5; definition of 225–6, 264–5; and emotional sensitivity 247; and feminist political ecology 39n8; and everyday knowledge 269; and life consciousness 34–5, *38*, *272*; and policy making 232–3, 248
experience-based approach (*keiken-ron*) 32, 144, 154n2, 225, 261; *see also* experientialism
experientialism (*keiken shugi*) 32, 144, 261, 264–5
experientialist approach *see* experience-based approach

famine food 146–7, 152–3
farming village urbanization policy (China) 105, 115
Fei, Xiaotong 107–8
feminist political ecology *13–4*, 39n8, 207, 211
feng shui 113
fishery 134–5, 272–3; aquaculture 137–8; Kamogawa Fisheries Cooperative 62, 66, 69; set net fishing 61–2, *63*, 69
folklore studies (*minzokugaku*) *see* Japanese folklore studies
foreseeability 201–2, 218
France 158, 217, 226, 281n20
Fukushima Daiichi Nuclear Power Plant 142, *143*, 153
Furukawa, Akira 144, 149, 154n1, 236, 244, 246

Germany 11, 280
ghosts *see* apparitions
Global South 11, 119, 191–2, 202
Gōdohara alluvial fan area *78*
Goffman, Erving 191
gong 105–6, 115–6; community 106, 115; personified 108, 115; and *si* 106–8

Great East Japan Earthquake Disaster 131, 133–4, 137, 152, 154, 156, 158–9, 245
Great Tōhoku Earthquake *see* Great East Japan Earthquake Disaster
Guandimiao 108–10, *110*
Guan Yu 110, 112
Guha, Ramachandra 15
Gujo City, Gifu 170
Gujo Odori (Gujo Dance) 170–3; Ohayashi Club 178–9; Preservation Society 173–4

hama-barai (ura-barai) 139
Harie District, Shiga 50
Harn, Lafcadio 21
Hasegawa, Koichi 225–6, 241, 245, 249n7
Hayashi, Chikio 279n2
historic environment 169–70, 182
Hobsbawm, Eric 170
Home, Robert 181–2
Horikawa, Saburō 249n3

industrialization 2, 11, 16, 77, 186, 258
iibun 21, 32, 35–8, 76, 93, 215, 228, 242–3, 277–9, 281n18, 281n20
ideal-typus (ideal types) 257
India 4, 6, 16
industrialization 2, 11, 16, 77, 186, 258
Inokuchi, Shōji 176–7
Ishinomaki City, Miyagi 158, 163–4

Japan Communist Party 84
Japanese folklore studies (*nihon minzokugaku*) 2, 8, 19, 28, 179, 224–7, 241
JCO 92; adjacent areas 99, *100*; criticality accident (1999) 95–6; incinerator project 96
Jeoksan-Kaok (enemy's house) 181–2, 187–9, *188*
jichikai 37, 51–3, 76, 81–4, 97, 99, 102, 103n3, 171–2, 260, 268, 277
joint right of possession (*kyōdō senyū ken*) 107

kabata 50–1, *52*, 214, 216–7
Kada, Yukiko 236
Kaino, Michitaka 107
Kamogawa City, Chiba 58
Kamogawa Marina Development Project (KMDP) 64–5
Kankyō Mondai no Shakai Riron (Sociological Theory of Environmental Problems) 3, 255
Karakuwa (peninsula) 134, 136

Index

kasenjiki (riverbed area) 118
kawabata 122
Kawauchi Village, Fukushima 142, 144–5
keiken-ron see experience-based approach
Kikuchi, Isao 146
knowledge: environmental 211; gendered 77; indigenous 77, 216–7, 237; life 86n2, 243, 248; modern(ist) 144, 238; situational 211; traditional 77, 86, 216; everyday *see* everyday knowledge; expedient *see* expedient knowledge
kōi-ron see action-based research
Koizumi, Yakumo *see* Harn, Lafcadio
kokoro 19–20, 21n4, 226–7
Korean War 183
Kōza school 269, 281n12
kuchi-kiki 68, 70–2
kuchiyose 163
Kuhn, Thomas 240
Kyoto 2, 47, 257

Lake Biwa 2–3, 26–7, 50, 224, 235–6, 244–5, 255–7
Latin America 4–5, 212
Lefebvre, Henri 280n11
life-centered approach (*seikatsu-ron*) xv–xvi, 28, 60, 76
life consciousness 32, 34–5, 53, 55, 267, 269, 271, 280n10, 281n14; Kiazaemon Aruga's idea of 271
life organization 20, 32, 37–8, 39n9, 59, 68–70, 214, 269–77, 281n17
life-environmentalism: and commons 118, 149; conceptual framework of *38*, *272*, *276*; criticisms against 225–6, 240–1, 244–5; and emotion 33–4, 218–9, 246–7, 260–1; and environmental justice *13–4*, 207, 214–5, 243–4; and experience (*keiken*) 32–3, 144, 225–6, 242, 262–4; and historic environment 169, 181–2; and knowledge 216–8, 240, 242–3, 269; and Japanese folklore studies 19–20, 226–7, 241; and life consciousness (*seikatsu ishiki*) 34–5, 271; and the logic of persuasion/justification (*iibun*) 35–6, 93, 242–3, 277–8; limitations of 4; and *minjian* 10; origins of 2, 25–6, 28, 224–5, 235–6; and policy actions 26–7, 228–9, 232–3, 237–8, 245, 248; and political ecology *13–4*, 207; positioning of *26*; and power 37, 215–6, 229, 267; and sustainable development 230–1, 258–9; and weak theory 19
life-world 266–7, 269

living common sense (*seikatsu jōshiki*) 35, 38, 217, 269, 271–4, 278, 281n20
living deceased *162*, 162–4
logic of justification *see iibun*
logic of persuasion *see iibun*

Manila (the Philippines) 192–3
Marxism/Marxist 28–9, 280n6, 281n12
Matsukawa Nature Conservation Society (MNCS) 79–80
Matsukawa Nature Conservation Society (MNCS) 79–81
Matsukawa Village, Nagano 77–9
medaka (*Oryzias latipes*) 238
Meiji: Great Municipal Merger of 85; period 9, 28, 39n7, 149, 153, 172, 182, 230; Restoration 7, 76, 149
mentalités 226
minjian 10, 17–8, 39n9, 75–6, 111
Miura, Kōkichirō 225, 241
Miyamoto, Tsuneichi 29
Mizoguchi, Yūzo 106, 108
Mizu to Hito no Kankyōshi (The Environmental History of Water and People) 3, 225, 257, 265
modern technocentrism (*kindai gijutsu shugi*) 25–7, 208–9, 232, 236–8, 256–7, 262
modernization 2, 4, 7, 9–10, 15, 35, 64, 75–6, 144, 170, 184; *see also* ecological modernization
Motoji, Matsuda 76–7, 86n2, 86n3, 93, 217, 236, 242–4, 249n4, 281n16
Motoori, Norinaga 28
Moune community in Kesennuma, Miyagi *135*
Mount Unzen Fugen, Nagasaki 216–7
multiculturalism 18, 244
multinaturalism 18
mura 30, 39n2, 76, 269–71, **273**
Murakami, Yōichirō 240
mutual non-understanding 33–4, 68, 245, 265, 280n9

Nakayama, Shigeru 240
Nash, Roderick 15
National Biodiversity Strategy and Action Plan (Japan) 238
National Environmental Policy Act (USA) 213
National Model Rural Village 115
Native Americans 157, 164, 212
natural environmentalism (*shizen kankyō shugi*) 25–7, 208, 232, 236–8, 256–7, 262, 280

Index 287

neighborhood association *see jichikai*
nichijo-chi see everyday knowledge
Nishihara District, Nagano 79–83, 86
nongovernmental organization (NGO) 15, 39n9, 75–6, 141n10, 184–5
nuclear: facility 91–2, 97, 101, 245; fallout 1; host community/locality 91–3, 94–5; power plant 17, 34, 95, 152, 265; village 39

oki-dashi 135–6, 141n7
Okinawa 182, 281n13
Osaka 2, 257
Ostrom, Elinor 49, 119
Our Common Future (1987) 230, 258–9
Ōura District (Kamogawa City) 61–3

Pacific Manufacturing Belt 2, 77
Philippines 6, 191
pokeweed (*Synurus pungens*) 147
political ecology 10–2, *13–4*, 18, 37, 207, 208–9, 212; *see also* feminist political ecology
pollution 208, 212, 249n3, 249n4; air 1, 17, 249n1; industrial 208, 238, 249n3; noise 65; tourism 47; water 2–3, 17, 80, 85, 224, 249n1
popular morality 35, 39n7, 269

Residents Demanding Safety of Operations and Peace of Mind 98–9
resilience 74–5, 140, 143–4, 146–7, 149, 153, 156–7, 164, 208
rice 51, 145–6, 148, 152, 197–8; cakes 147; crackers 126; *Jōkoku* variety 145; paddies 77, 82
River Act (Japan): the 1997 revision of 237

Sakuma, Girin 144–8
Sankyō Kaihatsu Co. 79–81, 83
Sanriku region 131, 133–4
Satō, Hiroo 163
satoyama 118, 120, 238, 249n2
secondary nature (*nijiteki shizen*) 238
Second World War 84, 173, 182, 183, 208, 268–9; pre- 174, 175, 225, 270; post- 2, 29, 107, 173
seikatsu 21, 27–9, 39n2, 236, 241, 262; *ishiki see* life consciousness; *jōshiki see* living common sense; right for the weak 149; and *seizon* (survival) 28; *soshiki see* life organization
seikatsu kankyō shugi see life-environmentalism

seikatsu-ron see life-centered approach
self-governing association *see jichikai*
Shimoda, Naoharu 266
Shutō, Toshikazu 108
Singapore 6, 193
sōgo murikai see mutual non-understanding
South Korea 5, 6, 181
spectrum thinking 94–5, 100–2
Surfrider Foundation Japan (SFJ) 66–8, 72
sustainability 1–3, 11, 28, 45, 55, 59–60, 119, 169, 219–20, 238–9
sustainable development 1–2, 230–1, 235, 258–9, 280n5
sustainable livelihoods xv, 3, *13–4*
Suzuki, Eitarō 271

taiken-chi (personal experiential knowledge) 269
Taiwan 16, 158
Teichi 62, 64, 66, 213; as a life organization 68–70; collective sense of 71–2; members 67, *71*; *see also* set net fishing
Tamanoi, Yoshirō 271
Temple Reconstruction Committee 111
Tenpo Famine 144–7
Thailand 5, 6, 191–2, 199
Three Kingdoms Period (China) 110
Three Power Sources Development Laws (*Dengen Sanpō*) 91
tianxia 106
Time and Space Research Institute 184
Tōkai Village, Ibaraki 95–6
Tokyo Electric Power Company (TEPCO) 153
Torigoe, Hiroyuki 2–3, 126, 192, 224, 236, 240, 255
total institution 191–2, 195, 202
tourism: aqua 45–8; coastal 213; cultural 169–70, 189; golf 77–9; historical 189; surfing 59–60
tsunami 39n8, 131, 213, 215, 218; Chile Earthquake-Tsunami (1960) 137; Indian Ocean Tsunami (2004) 213; Meiji-Sanriku-Ostunami (1896) 133, 134, 137; and oyster farming 138–9; Showa-Sanriku-Otsunami (1933) 133, 134, 137–8; in Tōhoku in March 2011 134–5, 143, 156–9
tsūzoku dōtoku see popular morality
two-persons world 37, 216, 269

Uchida, Tatsuru 160
United Kingdom (UK) 11, 158, 259

United Nations 28, 55, 235, 257; Conference on the Human Environment 258; Special Commission on the Environment 258; General Assembly 259; Environment Programme 280n4
United States (USA) 8, 11, 12, 20, 66, 158, 207–8, 213, 236, 240, 243
urbanization 2, 105–9, 113–5, 202, 217, 255

Wakamatsu, Eisuke 162–3
weak theory 19, 220

Weber, Max 32, 191, 260–1, 265
WWII *see* Second World War

Xintai City, Shandong Province 108
Xintai City Comprehensive Urban Plan (2004–2020) 109

yakata float *171*, 174, 177
Yamaguchi, Yaichirō 132
Yanagita, Kunio 8–9, 28, 179, 192, 225–7
Yasumaru, Yoshio 39n7

Printed in the United States
by Baker & Taylor Publisher Services